# Confronting Drought
# in Africa's Drylands

# Confronting Drought in Africa's Drylands

## Opportunities for Enhancing Resilience

Raffaello Cervigni and Michael Morris,
Editors

A copublication of Agence Française de Développement and the World Bank

**Library of Congress Cataloging-in-Publication Data**

Names: Cervigni, Raffaello, editor. | Morris, Michael L., editor.
Title: Confronting drought in Africa's drylands : opportunities for enhancing resilience / Raffaello Cervigni and Michael Morris, editors.
Other titles: Africa development forum.
Description: Washington, DC : World Bank : Agence Francaise de Development, 2016. | Series: Africa development forum series | "This book provides a synthesis of the study, 'The Economics of Resilience in the Drylands of Africa.' The study was a collaborative effort involving contributors from many organizations, working under the guidance of a team comprised of staff from the World Bank Group (WBG), the United Nations Food and Agriculture Organization (FAO), and the Consultative Group for International Agricultural Research Program on Policies, Institutions, and Markets (CGIAR-PIM)."
Identifiers: LCCN 2016010168 | ISBN 9781464808173
Subjects: LCSH: Arid regions--Africa, Sub-Saharan--Economic conditions--21st century. | Sustainable development--Africa, Sub-Saharan. | Drought management--Africa, Sub-Saharan. | Poverty--Government policy--Africa, Sub-Saharan. | Africa, Sub-Saharan--Economic conditions--21st century. | Africa, Sub-Saharan--Social policy.
Classification: LCC HC800.Z9 .E5369 2016 | DDC 338.967--dc23
LC record available at http://lccn.loc.gov/2016010168

# Africa Development Forum Series

The **Africa Development Forum Series** was created in 2009 to focus on issues of significant relevance to Sub-Saharan Africa's social and economic development. Its aim is both to record the state of the art on a specific topic and to contribute to ongoing local, regional, and global policy debates. It is designed specifically to provide practitioners, scholars, and students with the most up-to-date research results while highlighting the promise, challenges, and opportunities that exist on the continent.

The series is sponsored by Agence Française de Développement and the World Bank. The manuscripts chosen for publication represent the highest quality in each institution and have been selected for their relevance to the development agenda. Working together with a shared sense of mission and interdisciplinary purpose, the two institutions are committed to a common search for new insights and new ways of analyzing the development realities of the Sub-Saharan Africa region.

## Advisory Committee Members

*Agence Française de Développement*
**Gaël Giraud,** Executive Director, Research and Knowledge
**Mihoub Mezouaghi,** Deputy Director, Research and Knowledge
**Guillaume de Saint Phalle,** Head, Knowledge Management Division
**Cyrille Bellier,** Head, Research Division

*World Bank*
**Punam Chuhan-Pole,** Acting Chief Economist, Africa Region
**Markus P. Goldstein,** Lead Economist, Africa Region
**Stephen McGroarty,** Executive Editor, Publishing and Knowledge Division
**Carlos Rossel,** Publisher

## Sub-Saharan Africa

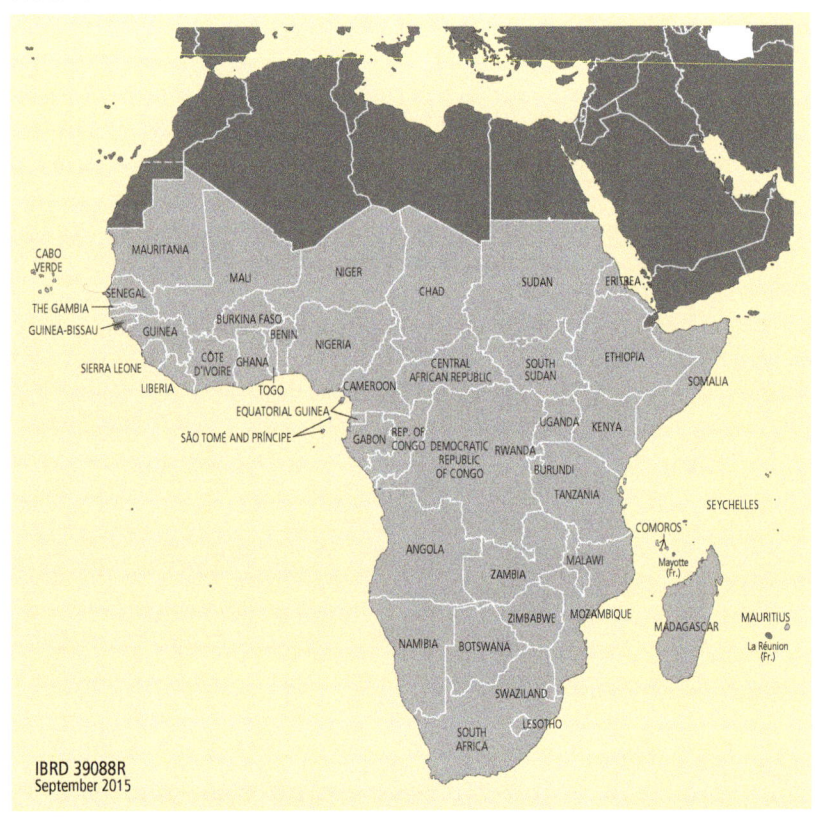

IBRD 39088R
September 2015

*Source:* World Bank (IBRD 390BBR, September 2015).

# Titles in the Africa Development Forum Series

*Africa's Infrastructure: A Time for Transformation* (2010) edited by Vivien Foster and Cecilia Briceño-Garmendia

*Gender Disparities in Africa's Labor Market* (2010) edited by Jorge Saba Arbache, Alexandre Kolev, and Ewa Filipiak

*Challenges for African Agriculture* (2010) edited by Jean-Claude Deveze

*Contemporary Migration to South Africa: A Regional Development Issue* (2011) edited by Aurelia Segatti and Loren Landau

*\*Light Manufacturing in Africa: Targeted Policies to Enhance Private Investment and Create Jobs*, «L'industrie légère en Afrique : Politiques ciblées pour susciter l'investissement privé et créer des emplois» (2012) by Hinh T. Dinh, Vincent Palmade, Vandana Chandra, and Frances Cossar

*\*Informal Sector in Francophone Africa: Firm Size, Productivity, and Institutions*, «Le système d'approvisionnement en terres dans les villes d'Afrique de l'Ouest : L'exemple de Bamako» (2012) by Nancy Benjamin and Ahmadou Aly Mbaye

*\*Financing Africa's Cities: The Imperative of Local Investment*, «Financer les villes d'Afrique : L'enjeu de l'investissement local» (2012) by Thierry Paulais

*\*Structural Transformation and Rural Change Revisited: Challenges for Late Developing Countries in a Globalizing World*, «Transformations rurales et développement : Les défi s du changement structurel dans un monde globalisé» (2012) by Bruno Losch, Sandrine Fréguin-Gresh, and Eric Thomas White

*The Political Economy of Decentralization in Sub-Saharan Africa: A New Implementation Model* (2013) edited by Bernard Dafflon and Thierry Madiès

*Empowering Women: Legal Rights and Economic Opportunities in Africa* (2013) by Mary Hallward-Driemeier and Tazeen Hasan

*Enterprising Women: Expanding Economic Opportunities in Africa* (2013) by Mary Hallward-Driemeier

\* Available in French

All books in the Africa Development Forum series are available for free at https://openknowledge.worldbank.org/handle/10986/2150

# Table of Contents

## Part C. Toward Policy Priorities

# Maps

# Tables

# Foreword

Drylands—defined here to include arid, semi-arid, and dry subhumid zones—
are at the core of Africa's development challenge. Drylands make up about 43
percent of the continent's land surface, account for about 75 percent of the area
used for agriculture, and are home to about 50 percent of the population,
including a disproportionate share of the poor. Due to complex interactions
among many different factors, vulnerability in drylands is high and is rising,
jeopardizing the long-term livelihood prospects for hundreds of millions of
people. Climate change, which is expected to increase the frequency and severity of extreme weather events, will exacerbate this challenge.

Most of the people living in the drylands depend on natural resource-based
livelihood activities, such as herding and farming, but the capability of these
activities to provide stable and adequate incomes is eroding. Rapid population
growth is putting pressure on a deteriorating resource base and creating conditions under which extreme weather events, unexpected spikes in global food
and fuel prices, or other exogenous shocks can easily precipitate full-blown
humanitarian crises and fuel violent social conflicts. Forced to address urgent
short-term needs, many households have resorted to unsustainable practices,
resulting in severe land degradation, water scarcity, and biodiversity loss.

African governments and their partners in the international development
community stand ready to tackle the challenges confronting drylands, but
important questions remain unanswered about how the task should be undertaken. Do dryland environments contain enough resources to generate the
food, jobs, and income needed to support sustainable livelihoods for a fast-
growing population? If not, can injections of external resources make up the
deficit? Or is the carrying capacity of drylands so limited that out-migration
should be encouraged?

To answer these questions, the World Bank teamed with a large coalition of
partners to prepare this book, which is designed to contribute to the ongoing
dialogue about measures to reduce the vulnerability and enhance the resilience

of populations living in drylands. Based on analysis of current and projected future drivers of vulnerability and resilience, the book identifies promising interventions, quantifies their likely costs and benefits, and describes the policy trade-offs that will need to be addressed when drylands development strategies are devised.

Sustainably developing the drylands and conferring resilience to their inhabitants will require addressing a complex web of economic, social, political, and environmental vulnerabilities. Good adaptive responses have the potential to generate new and better opportunities for many people, cushion the losses for others, and smooth the transition for all. Implementation of these responses will require effective and visionary leadership at all levels, from households to local organizations, national governments, and a coalition of development partners. This work, along with an accompanying series of background books, is intended to contribute to that effort.

*Makhtar Diop*
Vice President, Africa Region
The World Bank

# Acknowledgments

This volume is part of the African Regional Studies Program, an initiative of the Africa Region Vice Presidency at the World Bank. This series of studies aims to combine high levels of analytical rigor and policy relevance, and to apply them to various topics important for the social and economic development of Sub-Saharan Africa. Quality control and oversight are provided by the Office of the Chief Economist of the Africa Region.

This book reports the main findings and recommendations emerging from the study "The Economics of Resilience in the Drylands of Africa" (referred to henceforth as the Africa Drylands Study). The study was carried out as part of the World Bank's Africa Regional Studies Program under the direction of Shantaynan Devarajan, Francisco H.G. Ferreira, and Punam Puhan-Chole, with strong backing from Makhtar Diop.

The Africa Drylands Study was carried out as a collaborative effort and involved contributors from many partner organizations, working under the guidance of a team from the World Bank Group (WBG), the United Nations Food and Agriculture Organization (FAO), and the Consultative Group for International Agricultural Research Program on Policies, Institutions, and Markets (CGIAR-PIM) hosted by the International Food Policy Research Institute (IFPRI). WBG staff who participated in the coordination of the study included Raffaello Cervigni and Michael Morris (team leaders), with assistance from Paola Agostini and working under the direction of Magda Lovei. FAO staff who contributed to the coordination of the study included Mohamed Manssouri, Julia Seevinck, Pierre Gerber, and Anne Mottet. IFPRI and CGIAR-PIM staff who participated in the coordination of the study included Siwa Msangi and Karen Brooks.

This book draws on a series of thematic background papers prepared by the following authors:

**Drylands classification:** Guo Zhe and Jawoo Koo (IFPRI-PIM).

**Livestock:** Cornelis de Haan (World Bank consultant); Tim Robinson and Polly Ericksen (ILRI); Abdrahmane Wane, Ibra Toure, Alexandre Ickowicz, and Matthieu Lesnoff (CIRAD); Frederic Ham and Erwann Filliol (Accion Contre la Faim); Siwa Msangi (IFPRI); Pierre Gerber, Giulia Conchedda, and Anne Mottet (FAO); and Raffaello Cervigni and Michael Morris (World Bank).

**Agriculture Water Management:** Christopher Ward with Rafael Torquebiau (World Bank consultants) and Hua Xie (IFPRI).

**Irrigation Development:** Hua Xie, Weston Anderson, Nikos Perez, Claudia Ringler, Liang You, and Nicola Cenacchi (IFPRI).

**Agriculture:** Tom Walker (World Bank); Tom Hash, Fred Rattunde, and Eva Weltzien (ICRISAT); Jawoo Koo (IFPRI); Federica Carfagna (WFP); and Raffaello Cervigni and Michael Morris (World Bank).

**Tree-based Systems:** Frank Place and Dennis Garrity (ICRAF) and Paola Agostini (World Bank).

**Landscape Approaches:** Erin Gray, Norbert Henninger, Chris Reij, and Robert Winterbottom (WRI) and Paola Agostini (World Bank).

**Vulnerability and Resilience:** Pasquale Scandizzo, Sara Savastano, and Adriana Paolantonio (University of Rome); and Alberto Zezza and Marco D'Errico (World Bank).

**Social Protection:** Carlo del Ninno, Sarah Coll-Black, and Pierre Fallavier (World Bank).

**Human, Social, and Political Dimensions of Resilience:** Carol Kerven and Roy Behnke (Odessa Centre); Mohamed Manssouri, Julia Seevinck, AnnaLisa Noack, and Ahmed Sidahmed (FAO); Abdrahmane Wane, Ibra Toure, and Alexandre Ickowicz (CIRAD); Roger Blench (Mallam Dendo, Ltd.); Hamath Amadou Dia (Assane Seck Ziguinchor University); Katherine Homewood (University College London); Peter Little (Emory University); John McPeak (Syracuse University); Mark Moritz (Ohio State University); Michael Mortimore (Bayero University); and John Morton (Natural Resources Institute).

**Markets and Trade:** John Nash, Paul Brenton, and Alvaro Federico Barra (World Bank).

**Disaster Risk Management:** Carl Christian Dingel, Christoph Putsch, Vladimir Tsirkunov, Jean Baptiste Migraine, Julie Dana, and Felix Lung (World Bank).

**Land Degradation:** Riccardo Biancalani, Monica Petri, and Sally Bunning (FAO).

**Vulnerability Modeling:** Federica Carfagna (WFP), Joanna Syroka, Balthazar Debrouwer, and Elke Verbeeten (ARC); and Pierre Fallavier and Raffaello Cervigni (World Bank).

Many other partners and stakeholders active in drylands development efforts contributed to the study by participating in meetings and workshops, providing data and other research materials, or commenting on emerging findings and preliminary results: Severin Kodderitzsch, Martien van Nieuwkoop, Laurent Msellati, Benoit Bosquet, Stephen Danyo, Madjiguene Seck, Jacob Burke, Francois Onimus, Pierrick Fraval, Francois Le Gall, Andrew Dabalen, Ruth Hill, and Donald Larson (World Bank); Ahmed Sidiahmed and Dominique Burgeon (FAO); Djime Adoum and Edwige Botoni (CILSS); and Mahboub Malim (IGAD). Peer reviewers included Marianne Faye, Carter Brandon, and Stephen Mink (World Bank), as well as an anonymous external peer reviewer.

Administrative and logistical support was provided by Mapi Buitano, Nevena Ilieva, Marie Bernadette Darang, Jayne Kwengwere, Virginie Vaselopulos, and Mark Green (World Bank); Andrea LoBianco (FAO); and Gayane Markaryan (IFPRI).

Preparation of this book was coordinated by a small team led by Raffaello Cervigni and Michael Morris (World Bank) and including Elizabeth Minchew, Valerie Ziobro, Luis Liceaga, Vanthana Jayaraj, and Amy Gautam (World Bank Consultants). The publication process was managed by Stephen McGroarty, Abdia Mohamed, Aziz Gökdemir, and Andrés Meneses (World Bank). Michael Alwan (World Bank Consultant) proofread the book and revised the interior and the cover.

The generous financial support of the following partners is gratefully acknowledged: the Nordic Development Fund (which provided financial and technical assistance in particular for the Livestock background paper); the European Union and the Netherlands Ministry of Foreign Affairs (through their support of the TerrAfrica Leveraging Fund), the United Nations Food and Agriculture Organization (FAO), the Program on Forests (PROFOR) Trust Fund, and the Research Program on Policies Institutions, and Markets (PIM) of the Consultative Group for International Agricultural Research (CGIAR).

## Africa Drylands Study Collaborators

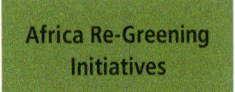

Africa Re-Greening Initiatives

African Risk Capacity

Agricultural Research for Development

CGIAR Research Program on Policies, Institutions, and Markets

Food and Agriculture Organization of the United Nations

Intergovernmental Authority on Development

International Crops Research Institute for the Semi-Arid Tropics

International Food Policy Research Institute

International Livestock Research Institute

Permanent Interstates Committee for Drought Control in the Sahel

World Agroforestry Center

World Resources Institute

## Africa Drylands Study Financial Contributors

CGIAR Research Program on Policies, Institutions, and Markets

European Union

Food and Agriculture Organization of the United Nations

Netherlands Ministry of Foreign Affairs

Program for Forests

TerrAfrica Leveraging Fund

# About the Editors

**Raffaello Cervigni** is a lead environmental economist with the Africa Region of the World Bank. He holds master's and PhD degrees in Economics from Oxford University and University College London. He has 20 years of professional experience in programs, projects, and research financed by the World Bank, the Global Environment Facility, the European Union, and the Government of Italy in a variety of sectors. He is currently the World Bank's regional coordinator for climate change in the Africa Region, after serving for about three years in a similar role for the Middle East and North Africa Region. He is the author or co-author of more than 40 technical papers and publications, including books, book chapters, and articles in academic journals.

**Michael Morris** is a lead agriculture economist with the Agriculture Global Practice of the World Bank. He holds master's and PhD degrees in Agricultural Economics from Michigan State University. He has co-authored World Bank flagship publications on fertilizer policy and agricultural commercialization, and he contributed to the *World Development Report 2008: Agriculture for Development*. His areas of expertise include agricultural policy, farm-level productivity enhancement, marketing systems and value chain development, agricultural research and technology transfer, innovation systems support, institutional strengthening, and capacity building. Prior to joining the World Bank, he spent 16 years in Mexico, Thailand, and Washington, DC with the International Maize and Wheat Improvement Center (CIMMYT) and the International Food Policy Research Institute (IFPRI).

# About the Authors

**Paola Agostini** is a lead environmental economist in the World Bank Environment and Natural Resources Global Practice. She is currently the global lead for Resilient Landscapes, where she examines projects and programs that try to improve the connectivity of protected areas, forests, agroforestry, rangeland, and agricultural land so as to increase productivity, community resilience, and production of ecosystem services. She holds a PhD in Economics from the University of California, San Diego, and a master's degree in Economic and Social Sciences from Università Bocconi, Milan, Italy.

**Paul Brenton** is a lead economist in the Trade and Competitiveness Global Practice of the World Bank. He is co-editor of the book *De-Fragmenting Africa: Deepening Regional Trade Integration in Goods and Services*, as well as the World Bank report, *Women and Trade in Africa: Realizing the Potential*. Dr. Brenton joined the World Bank in 2002, having been senior research fellow and head of the Trade Policy Unit at the Centre for European Policy Studies in Brussels. Before that he lectured in economics at the University of Birmingham, U.K. He holds a PhD in Economics from the University of East Anglia, U.K. A collection of his work was recently published in the World Bank's World Scientific Studies in International Economics volume, *International Trade, Distribution and Development: Empirical Studies of Trade Policies*.

**Federica Carfagna** is a vulnerability analyst for African Risk Capacity, having been with ARC since its inception in 2009. She is one of the main authors of the methodology used within Africa RiskView, the technical engine of ARC, to calculate food security in vulnerable and drought-affected populations. Ms. Carfagna holds a master's degree in Statistics from the University of Rome, "La Sapienza," and spent one year in an exchange program at the Cass Business School in London. Prior to joining ARC, she worked as statistician for WFP, for UN-DESA in New York, IFAD and Rome City Hall, and also as a data analyst to WFP's "World Hunger Series" and many school feeding publications.

**Giulia Conchedda** is a consultant geospatial data analyst with the Food and Agriculture Organization of the United Nations. She holds master's degrees in

Tropical Agriculture and Remote Sensing Tools applied to the monitoring of natural resources, and received a PhD in Geography from the Université Catholique of Louvain-la-Neuve, Belgium. She has some 15 years of professional experience in projects and research as a geospatial analyst with FAO, the World Food Programme, the Joint Research Centre of the European Commission, the International Livestock Research Institute of CGIAR, and the National Wetlands Research Center of the U.S. Geological Survey. She is coauthor of pathbreaking analyses and spatial modeling efforts on the distribution of livestock species and livestock production systems.

**Cornelis (Cees) de Haan** is a retired World Bank senior advisor. He holds a post-graduate degree in Animal Production from the University of Wageningen, Netherlands. He worked for 10 years in Dutch technical assistance programs in rural development projects in South America, followed by seven years in livestock research at the International Livestock Center for Africa (now ILRI) in Addis Ababa, Ethiopia, where he became deputy director general. He joined the World Bank in 1983, working for 10 years on livestock development in the African and East European regions, and for the same period as advisor responsible for policy development and quality enhancement in the World Bank's animal resource development activities in the Rural Development Department. He has contributed to World Bank policy and investments in livestock-related environmental, health, and social issues, and has published extensively in those fields. Since his retirement in 2001, he has remained active as a consultant in animal production and health for the World Bank and other international organizations.

**Carlo del Ninno** is a senior economist in the Africa Region of the World Bank working on safety net policies and programs. He holds a PhD from the University of Minnesota and has published on safety nets, food policy, and food security. He is currently the manager of the Sahel Adaptive Social Protection Program. Over the past 13 years, he has worked on analytical and operational issues on safety net programs covering several countries in South Asia and Sub-Saharan Africa. Before joining the World Bank, he worked on food security policy for the International Food Policy Research Institute in Bangladesh, and on poverty analysis in several countries for the Policy Research Division of the World Bank and Cornell University.

**Carl Christian Dingel** is a disaster risk management specialist with the World Bank. He holds a master's degree in International Land and Water Management from Wageningen University, Netherlands, and an engineering degree (Diplom Ingenieur) from Osnabrück University of Applied Sciences, Germany. He has 10 years of professional experience in disaster risk management and water, land, and natural resource management in Africa, South Asia, and Europe. He has led and co-led a number of disaster risk reduction projects, supported post-disaster assessments following floods and droughts across

Africa, and contributed to the 2011 Horn of Africa and 2012 Sahel drought recovery programs. He previously worked in applied research and development projects for German Technical Cooperation, the International Center for Agricultural Research in Dry Areas, as well as consulting firms and government agencies in Germany and the Netherlands.

**Polly Ericksen** leads the Livestock Systems and Environment Program at the International Livestock Research Institute (ILRI) in Nairobi, Kenya. She earned a master's in Economics and a PhD in Soil Science, both from University of Wisconsin–Madison. Her areas of expertise are adapting food systems to global environmental change to enhance both food security and key ecosystem services; researching options for lessening the vulnerability of pastoral livelihoods; and developing strategies for adaptation to climate change in agricultural systems. Prior to joining ILRI she worked for the ASB System-wide Program of CGIAR; Catholic Relief Services; the International Research Institute for Climate and Society at Columbia University, New York; and the Environmental Change Institute at the University of Oxford, U.K. She has worked extensively in Latin America, Africa, and South Asia.

**Pierre Fallavier** is a planner and social scientist with 19 years of experience in development and humanitarian programs and policies in Asia and Africa, working with the World Bank and UN agencies, local governments, civil society, and academia. He specializes in community-based development and social protection in post-conflict and fragile states. During the last five years Dr. Fallavier has been working on linkages between disaster-risk reduction, social protection, and humanitarian response in countries particularly affected by the impacts of climate change. He holds a PhD in Urban and Regional Planning from the Massachusetts Institute of Technology and a master's in Community Planning from the University of British Columbia. He is currently Chief of Social Policy, Planning, Monitoring and Evaluation for UNICEF in South Sudan.

**Pierre Gerber** is a staff member of the Food and Agriculture Organization of the United Nations (FAO), currently serving as senior livestock specialist at the World Bank. He holds a PhD in Agricultural Economics from the Swiss Federal Institute of Technology in Zurich and master's degrees in Agronomy and Environmental Law from the Ecole Nationale Supérieure Agronomique de Rennes and University of Nantes, France. He has worked for more than 15 years analyzing trends in global livestock systems and their interactions with the environment. He coordinates a global program of work including analytical studies, partnerships, and field projects on issues including climate smart agriculture, metrics of sustainability, policy formulation, and efficiency of natural resource use in agricultural systems. He has authored more than 50 FAO reports, book chapters, and scientific papers on livestock, climate change, and natural resources.

**Zhe Guo** is a senior Geographic Information System (GIS) coordinator with the Environment and Production Technology Division of the International Food Policy Research Institute (IFPRI). His interests include spatial modeling, spatial statistics, data mining, and remote sensing and land classification. He has worked on multiple projects funded by the Gates Foundation, USAID, and the World Bank. Zhe Guo earned an M.S. in Natural Resource Science and an M.A. in geography in University of Maryland, College Park.

**Frédéric Ham** is an expert in geographic information systems (GIS) and disaster risk reduction (DRR). He holds a master's in GIS from Lund University and a bachelor's in Environmental Engineering from Strasbourg University, France. With more than 10 years of experience with international humanitarian organizations, including Action Contre la Faim (ACF), the Red Cross, and Doctors Without Borders (MSF), he has been responsible for the design of several GIS-based applications aimed at reducing the impact of natural disasters. In particular, he has been extensively involved in the development of remote-sensing-based early warning and surveillance systems covering the Sahelian pastoral regions. These developments led to the production of functional and recognized tools to address food insecurity and vulnerability to drought in these areas.

**Norbert Henninger** is a senior associate at the World Resources Institute (WRI) working at the intersection of poverty reduction, natural resources management, and governance. He holds a master's in Environmental Sciences from Johns Hopkins University and an MBA from the University of Mannheim, Germany. His work focuses on creating better information and tools to formulate and evaluate development cooperation programs, advance green growth strategies, and carry out environmental and social assessments. He has written technical reviews and publications on targeting agricultural research and poverty reduction programs, environmental and agricultural indicators, and integrated assessments of ecosystems and human well-being.

**Alexandre Ickowicz** is a veterinarian and research fellow at the International Cooperation Center in Agricultural Research for Development in France, specializing in animal production science in the tropics. He holds master's and PhD degrees in veterinary science, tropical animal production, and environmental science from Paris XII University. He has 18 years of professional experience in dryland areas of west and central Africa, collaborating with national, regional, and international research and development institutions (NARS, CILSS, ILRI, FAO, World Bank) in improving knowledge on pastoral and agro-pastoral livestock production systems and contributing to development programs. He is currently director of a joint research unit between CIRAD, INRA, and SupAgro in Montpellier, France named SELMET (Livestock Systems Dynamics in Mediterranean Areas and the Tropics), which is committed to research in southern Europe, Africa, Southeast Asia, and Latin America.

**Jawoo Koo** is a research fellow at the Environment and Production Technology Division of the International Food Policy Research Institute (IFPRI). He holds master's and PhD degrees in Agricultural and Biological Engineering from the University of Florida. He has more than 10 years of experience in the development of a large-scale, spatially explicit crop system modeling framework and its application in Sub-Saharan Africa. He currently serves as the leader of IFPRI's Spatial Data and Analytics Theme. He is the author of more than 20 technical papers and publications, including books, book chapters, and articles in academic journals.

**Mohamed Manssouri** leads the UN Food and Agriculture Organization (FAO) Investment Center Service for Europe, Central Asia, Near East, North Africa, and Latin America and the Caribbean. He is an agricultural economist with a master's degree from AgroParisTech. Previously he coordinated FAO's "Renewed Commitment to a Hunger-Free Horn of Africa," leading the development of resilience-enhancing strategic plans and investment programs in the Horn of Africa. Prior to joining FAO, he was country program manager with the International Fund for Agricultural Development (IFAD) for 12 years, where he led the development of country investment strategies and programs in West and Central African countries. His areas of expertise include agricultural and rural development, food security, and poverty reduction, with a focus on policy and investment.

**Anne Mottet** is a livestock policy officer with the Food and Agriculture Organization of the United Nations. She holds a master's in Agricultural Development Economics from AgroParisTech and a PhD in Agrosystems and Ecosystems from National Polytechnic Institute of Toulouse. She has more than 10 years of experience in the livestock sector in Europe, Oceania, and Africa, in areas such as international trade and markets, policy assessments, and resource use efficiency and economics.

**John Nash** is lead economist in the Agriculture Global Practice in the World Bank Africa Region. He holds master's and PhD degrees in Economics from the University of Chicago. Since joining the World Bank in 1986, he has worked in five of the Bank's vice-presidencies. Prior to 1986 he was an assistant professor at Texas A&M University and an economic advisor to the chairman of the U.S. Federal Trade Commission. He has written numerous books, journal articles, and op-eds on agriculture, trade policy, climate change, and natural resource management.

**Frank Place** is a senior research fellow with the Policies, Institutions, and Markets Program (PIM) hosted by the International Food Policy Research Institute (IFPRI), where he leads research on technology adoption and impact assessment. He holds master's and PhD degrees in Economics from the University of Wisconsin–Madison. Prior to joining PIM, he worked for more than 15 years for the World Agroforestry Centre in Nairobi. He conducted

many studies related to policy constraints to and impacts of agroforestry practices. Earlier he also worked for the Land Tenure Center and the World Bank
conducting studies of indigenous tenure systems in Africa.

**Claudia Ringler** is deputy division director of the Environment and
Production Technology Division at the International Food Policy Research
Institute (IFPRI). She co-leads the institute's water research program and is also
a co-manager of the Managing Resource Variability and Competing Uses flagship of the CGIAR Research Program on Water, Lands, and Ecosystems (WLE).
Her research focuses on water resources management and agricultural and
natural resource policies for developing countries. Over the last 10 years she has
undertaken research on the impacts of climate change on developing country
agriculture and on appropriate adaptation and mitigation options. She has written more than 100 publications in the areas of water management, global food
and water security, natural resource constraints to global food production, and
the synergies of climate change adaptation and mitigation.

**Joanna Syroka** is the Director of Research and Development for the African
Risk Capacity. In these roles she oversaw the ARC design phase work program
and now leads the agency's technical and research work. Prior to joining ARC
she worked with the World Bank and UN World Food Programme to develop
tailored weather and commodity risk management products for agricultural
and humanitarian applications in Africa, Asia, and Central and South
America. Her work led to the first sovereign-level weather derivative products
in Africa and the early farmer weather insurance transactions in India. Earlier
she worked as a commodity derivatives analyst for one of the United
Kingdom's largest utility companies. She holds a PhD in Atmospheric Physics
from Imperial College, London.

**Ibra Touré** is a senior scientist with the French Agricultural Research Centre
for International Development (CIRAD) and holds a PhD in Geography from
the University of Nice (France). He has led research on pastoral topics in the
Sahel for more than 20 years and has authored many scientific and technical
articles. He is currently working under a joint contract with the Permanent
Interstates Committee for Drought Control in the Sahel (CILSS) in
Ouagadougou, Burkina Faso. He co-launched the joint research unit "Pole
Pastoralism and Drylands" (PPZS) in Senegal. His main research is in developing tools to better address and support the management of pastoral production
systems through the production of spatial knowledge, the design of accurate
indicators, and the capacity building of partners. He contributed to the formulation of the Regional Support Project Pastoralism in the Sahel (PRAPS) in 2013–
15, launched by the World Bank.

**Tom Walker** is an agricultural economist and holds a master's from the
University of Florida, as well as master's and PhD degrees from Stanford
University. Working with international agricultural research centers and

universities, he has over 30 years of overseas experience in South Asia, Latin America, and Sub-Saharan Africa. Relying heavily on longitudinal village studies and household panel surveys, he has written extensively on the economic development of dryland agriculture. Multiple interdisciplinary research investigations with biological and physical scientists have significantly enhanced his experience. In 2015 he edited a book, published by the Center for Biosciences and Agriculture International (CABI), that reports on the collaborative effort of over 200 biological and social scientists in documenting varietal change and the performance of crop improvement programs in 30 countries in Africa.

**Abdrahmane Wane** is a senior drylands economist with CIRAD, in joint appointment with the International Livestock Research Institute (ILRI) in Nairobi, Kenya. He holds master's and PhD degrees in Economics from the University Paris 9-Dauphine (France) focusing mainly on sovereign debt management. He was the coordinator of the joint research unit, "Pole Pastoralism and Drylands" (PPZS), in Senegal. His areas of expertise include development economics, cattle markets, price volatility and transmission, pastoral income distribution, food security, and value chain and network analysis, vulnerability/resilience. He is the author or co-author of more than 45 scientific publications including papers in peer-reviewed scientific journals, book chapters, and technical reports for leading institutions, and he has made at least 20 presentations at international conferences.

**Christopher Ward** is a research fellow in the Institute of Arab and Islamic Studies, University of Exeter. He holds degrees from Oxford University and is a Fellow of the Institute of Chartered Accountants in England and Wales. He worked for KPMG and McLintock Main Lafrentz in consultancy in the United Kingdom and the Middle East, and was Assistant Representative of the British Council in Saudi Arabia. He worked for 25 years in the World Bank. In the Africa Region, he focused on agriculture and irrigation, and lived in Kenya and Madagascar. In the Middle East and North Africa Region, he specialized in water and lived in Yemen and Morocco. He has authored numerous studies and papers, including the 2014 academic monograph "Water Crisis in Yemen."

**Hua Xie** is a research fellow at the International Food Policy Research Institute (IFPRI). He holds a PhD in environmental engineering from the University of Illinois at Urbana-Champaign. His area of expertise is water resources and environmental system analysis and modeling. At IFPRI, his research focuses on developing quantitative analytical and modeling tools to inform policy making for sustainable management of water and other natural resources key to agricultural development. Research topics of interest include: climate change impact on agricultural water resources, long-term projection of agricultural nutrient pollution, and evaluation of water land management technologies. He has been involved in a series of studies on irrigation investment potential in Sub-Saharan countries at both regional and national levels.

# Abbreviations

| | |
|---|---|
| ACF | Action Contre la Faim |
| ACMAD | African Centre of Meteorological Applications for Development |
| AGIR | Global Alliance for Resilience–Sahel and South Africa |
| AGRHYMET | AGRrometeorology, HYdrology, METeorology |
| AI | Aridity Index |
| ANR | assisted natural regeneration |
| ARC | African Risk Capacity |
| ARV | Africa RiskView model |
| AU | African Union |
| B/C | benefit/cost (assessment) |
| BAU | business as usual |
| BCR | benefit-cost ratio |
| BCSD | bias-correction spatial disaggregation |
| CGIAR-PIM | Consultative Group for International Agricultural Research Program on Policies, Institutions, and Markets |
| CIESIN | Columbia University Center for International Earth Science Information Network |
| CILSS | Permanent Interstates Committee for Drought Control in the Sahel |
| CIMMYT | International Maize and Wheat Improvement Center |
| CIP | International Potato Center |
| CIRAD | Agricultural Research for Development |
| CMIP5 | Coupled Model Intercomparison Project Phase 5 |
| COMESA | Common Market for Eastern and Southern Africa |
| DFID | Department for International Development, United Kingdom |
| DMP | Dry Matter Productivity |
| DRM | disaster risk management |
| DSSAT | Decision Support System for Agrotechnology Transfer |
| EAC | East African Community |
| ECOWAP | Economic Community of West Africa Agricultural Policy |
| ECOWAS | Economic Community of West African States |
| EU | European Union |

| | |
|---|---|
| FAO | United Nations Food and Agriculture Organization |
| FEWS NET | Famine Early Warning Systems Network |
| FMNR | farmer-managed natural regeneration |
| GAEZ | Global Agro-Ecological Zones database, FAO |
| GCM | Global Circulation Model |
| GEF | Global Environment Facility |
| GEPR | growth elasticity of poverty reduction |
| GFDRR | Global Facility for Disaster Reduction and Recovery |
| GHG | greenhouse gas |
| GIS | geographic information system |
| GLADIS | Global Land Degradation Information System |
| GLEAM | Global Livestock Environmental Assessment Model |
| GLW | Gridded Livestock of the World |
| GRUMP | Global-Urban Mapping Project |
| ha | hectare |
| HH | household |
| HSNP | Hunger Safety Net Program |
| ICARDA | International Center for Agricultural Research in the Dry Areas |
| ICPAC | Climate Prediction and Applications Center |
| ICRAF | World Agroforestry Centre (known as the International Centre for Research in Agroforestry [ICRAF] before 2002) |
| ICRISAT | International Crops Research Institute for the Semi-Arid Tropics |
| ICT | information and communications technology |
| IFAD | International Fund for Agricultural Development |
| IFPRI | International Food Policy Research Institute |
| IGAD | Intergovernmental Authority on Development, Africa |
| ILRI | International Livestock Research Institute |
| IMF | International Monetary Fund |
| IPCC | Intergovernmental Panel on Climate Change |
| IRR | internal rate of return |
| KfW | Kreditanstalt für Wiederaufbau (Reconstruction Credit Institute), Germany |
| km$^2$ | square kilometers |
| LADA | Land Degradation Assessment in Drylands Project |
| LEWS | Livestock Early Warning System |
| LSI | large-scale irrigation |
| LSMS | Living Standards Measurement Surveys |
| MT | metric ton |
| MV | modern variety |
| NARS | National Agricultural Research Systems |
| NDVI | Normalized Difference Vegetation Index |
| NGO | nongovernmental organization |

| | |
|---|---|
| NPV | net present value |
| NTM | non-tariff measure |
| OCHA | UN Office for the Coordination of Humanitarian Affairs |
| OIE | World Organisation for Animal Health |
| PDNA | Post Disaster Needs Assessment |
| PDSI | Palmer Drought Severity Index |
| PET | potential evapotranspiration |
| PPZS | Pole Pastoralism and Drylands |
| PRAPS | Regional Support Project Pastoralism in the Sahel |
| PROST | Pension Reform Options Simulation Toolkit |
| PSNP | Ethiopia Productive Safety Net Program |
| RCP | Representative Concentration Pathway |
| REC | Regional Economic Commission |
| SADC | Southern Africa Development Community |
| Safex | South African Futures Exchange |
| SHIP | Survey-based Harmonized Indicators Program, World Bank |
| SPAM | Spatial Production Allocation Model, IFPRI |
| SSI | small-scale irrigation |
| SSN | social safety net |
| TLU | Tropical Livestock Units |
| UNCCD | United Nations Convention to Combat Desertification |
| UN-DESA | United Nations Department of Economic and Social Affairs |
| UNDG | United Nations Development Group |
| UNDP | United Nations Development Programme |
| UNEP | United Nations Environment Programme |
| UNFPA | United Nations Population Fund |
| UNICEF | United Nations Children's Fund |
| USAID | U.S. Agency for International Development |
| WBG | World Bank Group |
| WFP | United Nations World Food Programme |
| WMO | World Meteorological Organization |
| WRI | World Resources Institute |
| WRSI | rainfall-based drought index |

# Overview

## The development challenge posed by drylands

Drylands—defined for purposes of this book based on the widely used Aridity Index[1] to include arid, semi-arid, and dry subhumid zones—account for three-quarters of Sub-Saharan Africa's cropland, two-thirds of cereal production, and four-fifths of livestock holdings. In East and West Africa—the focus of this book—drylands are home to over 300 million people, and they account for a large share of the poor, including many of those lacking access to basic services such as health care and education (map O.1).

Today frequent and severe shocks, especially droughts, limit the livelihood opportunities available to millions of households and undermine efforts to

**Map O.1** Dryland regions of West and East Africa

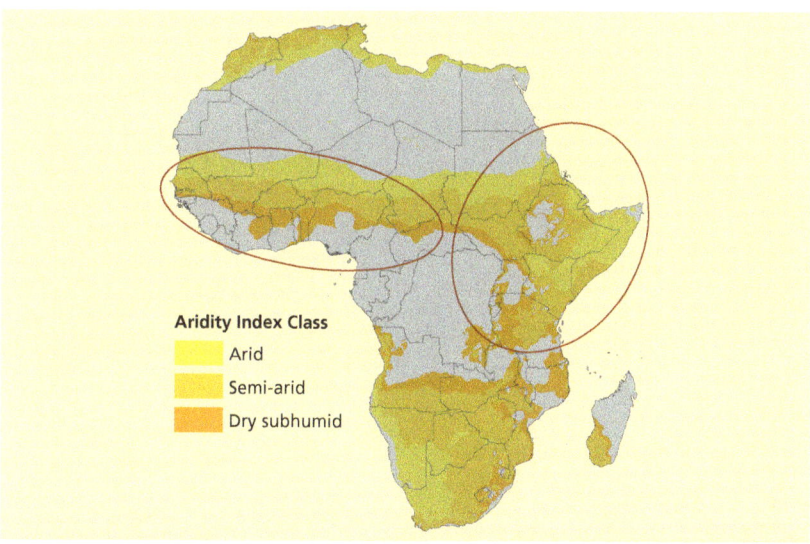

Aridity Index Class
- Arid
- Semi-arid
- Dry subhumid

*Source:* ©Harvest Choice, IFPRI, 2015. Reproduced, with permission from Zhe Guo; further permission required for reuse.

eradicate poverty in the drylands. These shocks regularly cause large drains on government budgets and consume a significant portion of the region's international development assistance, especially in the absence of robust social protection systems and rapidly scalable safety nets. As a result, scarce resources are diverted away from pursuing longer-term development goals and redirected to mobilizing costly short-term responses to humanitarian crises. In 2011 around US$4 billion was spent on humanitarian assistance to the Sahel and the Horn of Africa, equivalent to over 10 percent of total Official Development Assistance to all of Sub-Saharan Africa (OECD 2015). The challenges threatening the livelihoods of many of the groups that live in drylands are compounded by their social and political marginalization, which muffles their voices and limits their ability to influence political processes that affect their well-being.

If the current situation is precarious, the future promises to be even more challenging. By 2030 the number of people living in the drylands of East and West Africa is expected to increase by 65 to 80 percent (depending on the fertility scenario). Over the same period climate change could result in an expansion of the area classified as drylands, by as much as 20 percent under some scenarios, for the region as a whole, with much larger increases in some countries (map O.2). This would bring more people into an ever more challenging environment.

**Map O.2** Shift and expansion by 2050 of dryland areas due to climate change

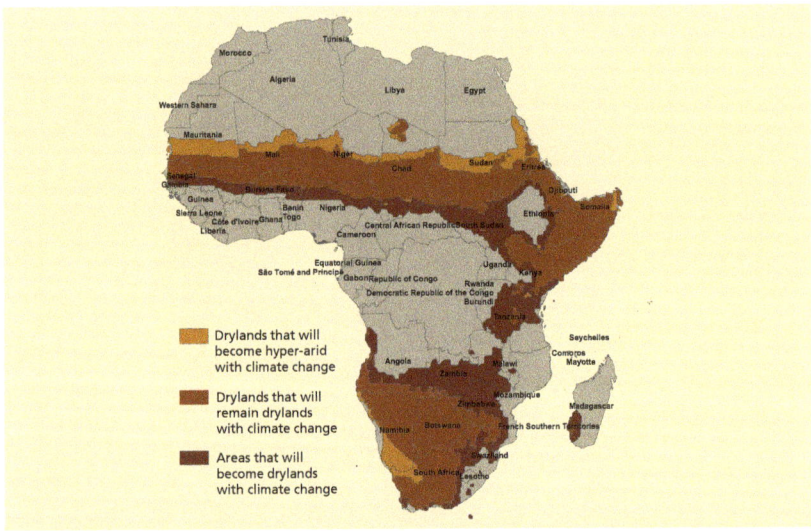

*Source:* Estimates based on Intergovernmental Panel on Climate Change (IPCC) data.

*Note:* The map shows the extent to which drylands (defined to include all zones with an aridity index 0.05–0.65) could shift and expand by 2050 as a result of climate change. To visualize the largest possible impacts, the map reflects the fastest growth of GHG (greenhouse gas) concentration (RCP 8.5 [Representative Concentration Pathways]) under the driest of a set of over 40 climate models.

## Scope of inquiry: Focus of this book

In response to a series of humanitarian crises, especially the drought-induced emergencies that occurred in the Horn of Africa in 2011 and the Sahel in 2012, national governments and the international development community have scaled up efforts to tackle the challenge of vulnerability in drylands through initiatives such as the Global Alliance for Resilience (AGIR)–Sahel and West Africa (facilitated by OECD) and the Global Alliance for Action for Drought Resilience and Growth (facilitated by the U.S. Agency for International Development [USAID]). These ongoing efforts are helping to address the recurring crises in the drylands, but the challenge is to ensure that the solutions they provide are not only temporary. Permanently reducing the vulnerability of the people living in drylands will require sustained efforts to attack the underlying root causes of their problems, using policies and programs that extend beyond relatively short electoral cycles. This book focuses on what should be the focus of the next generation of interventions aimed at enhancing the resilience of dryland populations in the face of demographic, economic, environmental, and climatic change.

If current trends continue, over the next two decades dryland regions of Africa will experience strong population growth. Higher population density in the drylands, combined with increasing interest from outside investors in large-scale commercial agriculture and extractive industries, will put additional pressure on the region's fragile natural resource base, pushing it in some cases beyond its regenerative capacity. As competition for resources intensifies, conflicts over land, water, and feed are likely to multiply, reducing the ability of governments, development agencies, and local communities to manage the impacts of droughts and other shocks.

In this context, building resilience to droughts and other shocks is of paramount importance. When households and communities are repeatedly hit by shocks and lack the means to respond, they then have difficulty accumulating the human, physical, and natural capital needed to lift themselves out of poverty. For this reason, building resilience to shocks is not necessarily a goal in itself, but remains an essential pre-condition for achieving higher-level development goals, such as poverty eradication, sustainable improvements in living conditions, and food security.

This book focuses on the medium-term prospects (over the next two decades) for increasing the resilience to drought and other shocks of people living in dryland areas of East and West Africa. Increasing resilience will not automatically lead to poverty eradication; for poverty to be eradicated, a number of additional actions will have to be taken, for example, improving health services, strengthening educational systems, and improving access to markets for inputs and outputs. But while increasing resilience is not a *sufficient* condition for poverty eradication, it is most likely a necessary one, because it is hard to imagine how households that are unable to cope with the impacts of drought and other shocks can save enough to augment their endowment of productive assets and increase their income generation potential.

The questions concerning vulnerability and resilience addressed in this book must be understood against the backdrop of an extremely dynamic environment. Dryland regions of Africa are already undergoing sweeping changes that are affecting the livelihoods of millions of households. Because the ongoing transformation of the drylands is being propelled by demographic drivers that have a great deal of momentum, the key question for policy makers is how best to manage the demographic, social, and economic changes that are coming.

Currently, most of the people living in the dryland regions of East and West Africa rely on herding and farming for their livelihoods. Over the longer run, structural transformation of the economy may generate opportunities for new livelihood activities that are less vulnerable to the impacts of droughts and other shocks. In the short to medium term, however, the key policy question concerns the extent to which current livelihoods can be made more resilient. In that context, this book examines two main areas of intervention, which are considered complementary.

1. **Improving current livelihood activities:** For the foreseeable future, most of the people living in drylands in East and West Africa will continue to make their living from herding and farming. For that reason, the book considers what can be done to make current livelihood activities more productive, more stable, and more sustainable, through investments supported by policy reforms and institutional change. The emphasis is on technological and management choices that have the potential to increase the returns from pastoralism, agro-pastoralism, and crop farming. Complementary activities in areas such as family planning, education, job creation, and financial markets are recognized as having a major influence on livelihood activities, but these complementary activities are not analyzed in detail.

2. **Strengthening social protection programs including safety nets:** In many parts of the drylands, even the most productive, stable, and sustainable livelihood activities will not be fully immune to the effects of droughts and other shocks. Households that rely on herding or farming as principal livelihood sources will continue to be exposed to droughts and other shocks, which depending on their frequency and severity can negatively affect incomes and plunge large numbers of people into poverty. For that reason, the book examines the degree to which social protection programs including safety nets can be used to strengthen the ability of dryland populations to cope effectively with the impacts of droughts and other shocks.

Improving current livelihood activities and strengthening social protection programs have significant potential to reduce vulnerability and enhance resilience of populations living in drylands, but both are likely to face limits, particularly in the face of technological, financial and fiscal constraints. In light of these limits, policy makers will need to consider a third set of interventions, namely, encouraging dryland populations to switch to alternative livelihood activities that are less vulnerable to droughts and other shocks. By assessing the scope and limitations of the first two types of interventions, this book helps define the importance across the group of countries analyzed of the third type of intervention. The book does not attempt to identify or analyze in detail the alternative livelihood activities that may offer the brightest prospects for dryland populations in East and West Africa; those tasks fall outside the scope of the present study and remain topics for future research.

Geographically the book focuses on dryland zones in East and West Africa, where vulnerability to drought and other shocks is highest. Many of the insights generated by the analysis have broader applicability, however.

## Conceptual framework: The determinants of resilience

Prospects for sustainable development of drylands are assessed in this book through the lens of resilience. But what exactly is meant by *resilience*? Most definitions of resilience relate to the ability of people or ecosystems, or both, to withstand and recover from shocks. In the context of drylands, the most important shocks are meteorological shocks, especially droughts, which are the main focus of the discussion that follows. Other shocks that are considered but not analyzed in detail include health shocks, price shocks, and conflict-related shocks.

In the absence of a single, widely accepted definition of resilience, this book uses a dimension-based approach (detailed in box O.1). Resilience—understood here to mean the ability of people to withstand and respond to droughts and other shocks—is affected by three types of factors:

- **Exposure** is the degree to which people are subject to droughts and other shocks, which depends mainly on where they live.
- **Sensitivity** is the degree to which people are affected by droughts and other shocks, which is determined by the nature and composition of their income sources and assets.
- **Coping capacity** is the ability of people to mitigate the impact of droughts and other shocks after they occur, through own resources, or support from friends, relatives, or the government.

---

**BOX 0.1**

## The dimensions of resilience

**Exposure** can be defined as the frequency and degree to which a household is subject to being hit by droughts and other shocks. A household whose assets are located in an area in which severe droughts occur once in every 5 years on average is more exposed than a household whose assets are located in an area in which severe droughts occur once in every 15 years on average. Exposure is an exogenous dimension of vulnerability, that is, outside the control of the household in the short run.

**Sensitivity** is the degree to which a household is affected by droughts and other shocks when they occur. For a given level of exposure, a household that derives a large share of its income from shock-affected activities (e.g., rainfed cropping and pasture-based livestock production) will have a higher sensitivity to the shock, other things equal, than a household that derives a small share of its income from shock-affected activities. Sensitivity is determined in large part by past decisions made by a household regarding the nature and mix of its assets (and by its livelihood strategy). Changing the nature and mix of assets (and the livelihood strategy) is one of the main avenues the household can follow to enhance its resilience.

**Coping capacity** refers to the ability of a household to mitigate the impact of droughts and other shocks after they occur. Access to financial resources (from its own savings, from friends or relatives, or from social safety nets) can help the household make up for an income shortfall resulting from, for example, a drop in production following a weather-induced shock. Liquidating productive assets to mitigate the negative impacts of current shocks may reduce the ability of the household to mitigate the impacts of future shocks, that is, it will reduce the household's resilience. Since it is unlikely that all risks can be avoided by diversifying household assets and altering income-generating activities to reduce exposure to future shocks, resilience-enhancing strategies usually consist of a combination of actions to reduce sensitivity to shocks and actions to increase coping capacity.

---

Other conditions being constant, the resilience of a household in the face of droughts and other shocks increases the lower its exposure, the lower its sensitivity, and the greater its coping capacity. Resilience is determined by the interplay of all three dimensions, so attempts to understand resilience in terms of just one or two dimensions can produce a misleading picture. For example, when relatively few people are living below the poverty line, it would be easy to conclude that the coping capacity of the population is relatively high, since most households have enough assets to be able to recover from a drought, should a drought occur. Based on such reasoning, policy makers might use the poverty headcount as an indicator of vulnerability. But focusing in this way on a single dimension of resilience could lead policy makers to overlook the fact that even though most households have enough assets to recover from a drought, the livelihood strategy that allowed them to accumulate those assets may be extremely sensitive to droughts. If this is the case, recurrent droughts could cause households to move in and out of poverty over time. In such a scenario, the population at risk should be understood to include not only the people that are poor today, but also the people who risk becoming poor tomorrow because their income is sensitive to droughts.

The importance of using a multidimensional approach to understand resilience can be seen by looking at the experience of several thousand Ethiopian households that participated in a series of surveys carried out during the period 1994–2009. Many of these households transitioned in and out of poverty, so during a period when the overall poverty headcount was gradually coming down, the fortunes of individual households were much more variable. On average, in any given year 16–17 percent of households started out poor and stayed poor, 18–19 percent of households started out non-poor and fell into poverty, 16–20 percent of households started out poor and climbed out of poverty, and 45–48 percent of households started out non-poor and remained non-poor (for details, see Scandizzo et al. 2014).

The Ethiopia household level evidence generates two important insights. First, policies that succeed in bringing some people out of poverty at a particular point in time do not necessarily guarantee that, as a result of subsequent shocks, many of those people will not fall back into poverty. Second, enhanced resilience is a pre-condition for sustained reduction and eventually eradication of poverty. As a result, it makes sense to explore policies and interventions that can increase resilience (as these will lay the foundation for poverty reduction); these policies and interventions should holistically address all three dimensions of resilience.

## Vulnerability in drylands if transformation is not managed

If current trends continue, how are patterns of vulnerability in African drylands likely to evolve? An original modeling framework developed for this book (referred to as the *umbrella model* because it integrates the results of more narrowly focused analyses carried out at the level of individual sectors) was used to assess the likely impacts of projected changes in the main drivers of resilience. The purpose of the umbrella modeling exercise was to assess the magnitude of the coming challenge and identify opportunities for policy interventions. The exercise generated a number of important insights, as follows.

**The number of people living in East and West Africa drylands who are exposed to droughts and other shocks will grow considerably.** In the absence of significant out-migration, by 2030 the population living in rural areas of the dryland countries is projected to grow by 15–100 percent (depending on the country).

**Economic growth will reduce the share of people living in drylands who are sensitive to droughts and other shocks, but probably not fast enough to overcome the effects of demographic growth.** As GDP growth generates new employment opportunities in the manufacturing and services sectors, the share of the population living in drylands and dependent on livestock-keeping and crop farming is likely to decrease. Nevertheless, in the presence of rapid population growth and increasing competition for resources from outside investors, the absolute number of people who depend on livestock-keeping and crop farming and who are exposed and sensitive to droughts and other shocks will likely outpace the exits out of agriculture. As a result, the total number of people dependent on agriculture is projected to increase (figure O.1).

**Economic growth will generate additional resources that can be used to cope with droughts and other shocks, but growth needs to become more pro-poor.** If GDP continues to grow in line with historical rates and the growth elasticity of poverty reduction averages 0.75 (a value that denotes relatively

**Figure O.1** Number of people in drylands projected to be dependent on agriculture in 2030 (2010=100, medium fertility scenario)

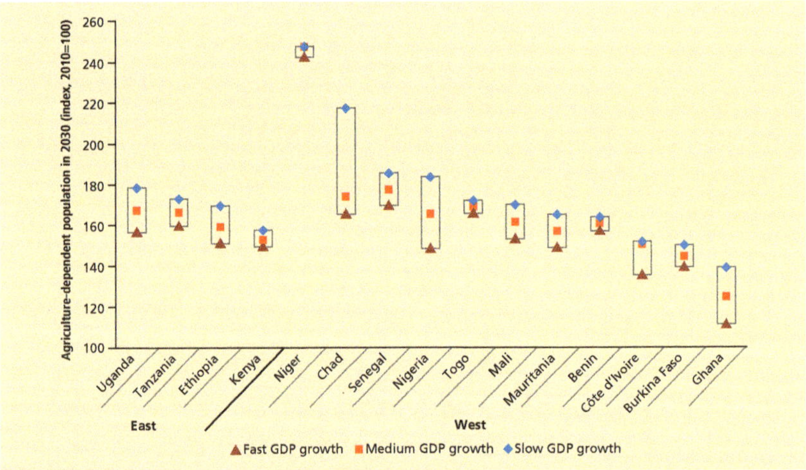

Source: Calculations based on World Bank data.

Note: The figures in the chart represent the number of dryland people projected to be dependent on agriculture in 2030 in relation to the corresponding figure in 2010. So for example, a figure of 140 indicates a 40 percent increase over the 2010 level of agricultural employment. For each country, the range is defined by different scenarios of per capita GDP growth, which is expected to generate some exit of employment out of agriculture as a result of structural transformation of the economy. The details of the calculation are provided in the appendix.

inequitable growth, such as that being observed in many African countries), the number of people in the drylands who depend on agriculture and live below the poverty line will increase in virtually every country (exceptions could include Burkina Faso in West Africa and Uganda in East Africa).

**Faster, more inclusive growth could reduce the incidence of vulnerability in drylands, but it will not eliminate vulnerability altogether.** Under an optimistic scenario that assumes that growth is both fast and equitable (unfortunately at odds with recent experience), the number of vulnerable people living in drylands could decrease by up to 40 percent in East Africa and up to 10 percent in West Africa (figure O.2). Despite these gains, the number of people needing assistance when droughts or other shocks occur is likely to exceed the reach of existing social protection systems, which suggests that large-scale humanitarian assistance would still be needed on a regular basis.

**Investment in the education of girls can help mitigate the size of the challenge, but it will not fully resolve the problem.** Investment in the education of girls has been shown to lower fertility rates over the medium to long term. As fertility rates fall, so does the number of people who are likely to need public assistance. The impact of reducing fertility rates, while non-negligible, is likely to be insufficient to address the problem. Using the UN low fertility population projections as a first-order approximation of the effects of fertility reduction

**Figure O.2** Vulnerable people in drylands in 2030 (2010=100, medium fertility scenario)

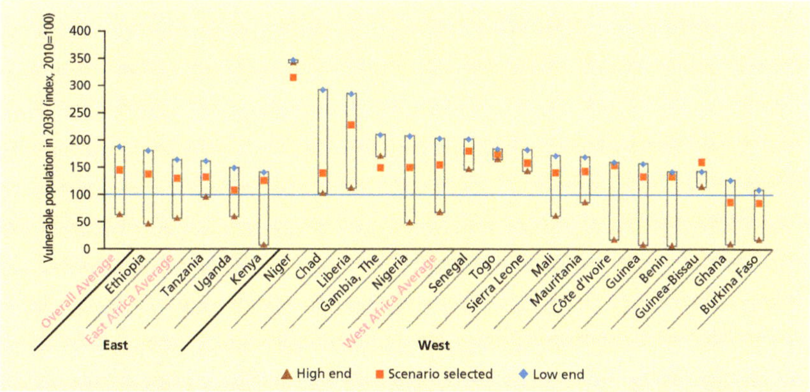

*Source:* Calculations based on World Bank data.

*Note:* The figures in the chart represent the number of dryland people projected to be employed in agriculture and having an income below the poverty line in 2030, compared to the corresponding number in 2010. For example, a value of 140 indicates a 40 percent increase by 2030 in the number of poor people employed in agriculture, compared to 2010. For each country, the range is defined under alternative scenarios involving different assumptions on per capita GDP growth and growth elasticity of poverty reduction. In particular, in the high-end scenario, growth rates and the income elasticity of poverty reduction are assumed to be at the 75th percentile of the distribution of the corresponding historical values. In the low-end scenario, they are assumed to be at the 25th percentile of the historical distribution. In the reference scenario selected, growth rates are set at the historical, country-specific average, while the growth elasticity of poverty reduction is set for all countries at the level of 0.75. Further details of the calculation appear in the appendix.

policies, the increase by 2030, compared to 2010, in the number of people vulnerable to droughts and other shocks could be reduced by a third (figure O.3).

## Options for increasing resilience

By 2030 economic growth leading to structural change will allow some of the people living in drylands to transition to non-agriculture-based livelihood strategies, reducing their vulnerability. Many other people living in drylands will continue to rely on livestock-keeping and crop farming. For the latter group, a number of best-bet interventions described in this book have the potential to make a significant difference in reducing vulnerability and increasing resilience. This book evaluates the key opportunities and challenges associated with these interventions, and it draws a number of conclusions that have important implications for policy making.

**Livestock-keepers in the drylands can be made more resilient through investments in improved management practices combined with support to new, complementary income sources.** Pastoralism and agro-pastoralism are the predominant forms of livestock-keeping throughout large parts of the drylands. Many pastoralists, particularly those at the lower end of the income spectrum, are vulnerable to falling into poverty (or sinking deeper into poverty) because their herds are not large enough to provide a reliable income stream in

**Figure O.3** Vulnerable people in drylands in 2030 (2010=100, different fertility scenarios)

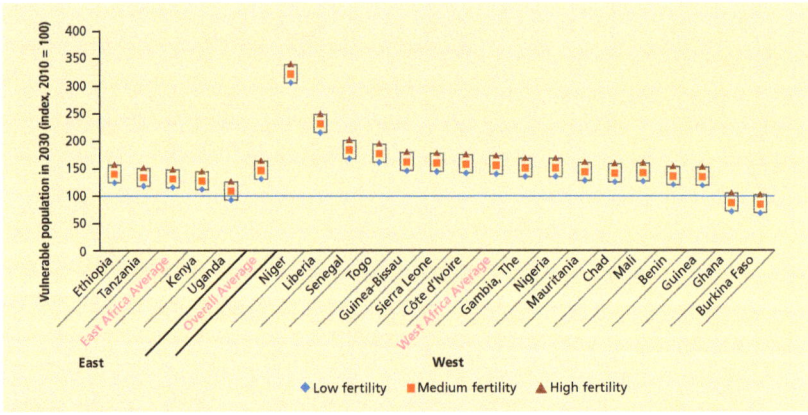

*Source:* Calculations based on World Bank data.

*Note:* The figures in the chart represent the number of dryland people projected to be employed in agriculture and having an income below the poverty line in 2030, compared to the corresponding number in 2010. For example, a value of 140 indicates a 40 percent increase by 2030 in the number of poor people employed in agriculture, compared to 2010. For each country, the range is defined by the three scenarios of population growth contained in the UN World Population Prospects (2012 Revision—http://www.un.org/en/development/desa/publications/world-population-prospects-the-2012-revision.html).

the face of erratic rainfall, recurring outbreaks of disease, continual conflict, and other shocks. In 2010 only 30 percent of households in the Sahel and the Horn of Africa possessed enough livestock assets to stay out of poverty in the face of recurrent droughts. With human population growth outstripping growth in livestock numbers, that share is projected to drop to 10 percent by 2030. Many livestock-keeping households (some 60 percent of the projected 2030 population) will feel pressure to drop out of livestock-based livelihoods, with the remaining 30 percent of households projected to stay in the system despite remaining vulnerable to droughts and other shocks.

Strategic interventions could slow the rate at which poor households feel pressure to abandon livestock-keeping, while at the same time boosting the income of those who remain. Productivity-enhancing interventions—such as providing improved animal health services, ensuring early offtake of young male animals, destocking quickly in the face of approaching drought, and ensuring improved access to grazing areas—could raise the share of resilient households by 50 percent (figure O.4). These gains would be achieved from a

**Figure O.4** Impact of improved animal health and early offtake of young bulls on the resilience status of livestock-dependent households in 2030 (% of households)

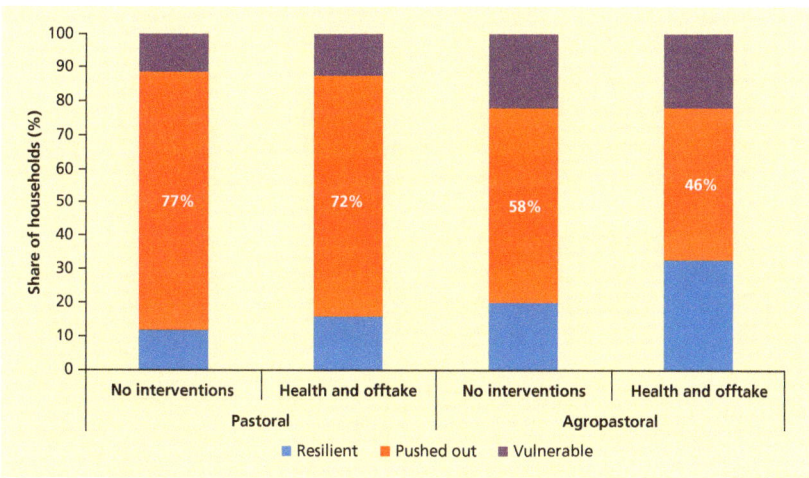

Source: De Haan et al. 2014.

Note: The figures in the chart represent the shares of households that are estimated to fall, without and with resilience interventions, into one of three categories. *Resilient households* are those that own herds above a resilience threshold required to withstand a sequence of high and low rainfall years similar to those experienced in the last 20 years. *Pushed-out households* are those that own herds below a lower survival threshold, so that they are unlikely to sustain themselves even in an average rainfall year. *Vulnerable households* are those whose herd size falls between the survival threshold and the resilience threshold. These households own enough animals to remain above the poverty line in an average year, but not enough to cope effectively during drought years. The figure refers to aggregated results for Ethiopia, Kenya, Uganda, Senegal, Nigeria, Mauritania, Chad, Mali, Burkina Faso, and Niger.

low starting point, however, so a large share of households would still remain vulnerable or feel pressure to drop out of livestock-keeping altogether (85 percent in pastoral areas, 70 percent in agro-pastoral areas). All told, more than 3 million households in 10 dryland countries could become resilient thanks to these interventions, at a cost of US$0.5 billion per year, or US$160 per household made resilient.

The scope for productivity-enhancing investments to increase livestock production in the drylands is limited by constraints on feed availability and by the rate at which animals can reproduce. Still, resilience of livestock-keeping households could be increased by interventions falling outside the domain of conventional livestock improvement programs—for example, policies designed to bring about a more equitable distribution of livestock assets: these could take the form of subsidized credit to enable smallholders to reach a minimum herd size, or progressive taxation of wealthier livestock owners. Some of these measures are prone to abuse, however (e.g., preferential credit programs), and others are likely to generate opposition from powerful groups with vested interests (e.g., progressive taxation regimes). If the potential disadvantages limit the scope for implementation, then it will be important to identify interventions that provide new income sources for poor livestock-keepers, such as programs that provide payments for environmental services. This will help limit exits from livestock-keeping and reduce the likelihood that those who continue to rely on livestock-keeping as their principal livelihood source will remain poor and vulnerable to shocks.

**Improved crop production technologies can deliver sizeable resilience benefits by boosting productivity in rainfed agriculture.** By 2030, if no action is taken, the number of farming-dependent households in the Sahel and the Horn of Africa that are poor and vulnerable to droughts and other shocks is projected to increase by around 60 percent. Interventions designed to improve the productivity of rainfed crops have the potential to dampen that increase considerably. Simulations of the impacts of the best-bet crop-intensification technologies (e.g., drought- and heat-tolerant varieties, improved fertility management, rainwater harvesting) on the productivity of key staples grown in drylands (maize, sorghum, and millet) suggest that the number of drought-affected poor households could be reduced by 10–80 percent compared to a "business as usual" (BAU) scenario, depending on the country and aridity zone. To ensure adoption, governments will need to address the technical, institutional, and financial challenges associated with the deployment of the best-bet technologies.

**Adding trees to current farming systems can further increase resilience.** Trees can improve the productivity and stability of crop and livestock production systems by providing multiple benefits that tend to stand up well in the face of weather shocks. Tree-based systems include systems based on farmer management of naturally occurring species (generally more appropriate in drier zones), as well as systems involving deliberate planting of economically useful species (generally more appropriate in more humid zones). When farmer-managed natural regeneration of native species is combined with the other productivity-enhancing technologies discussed in this book, the impact is impressive—the projected number of poor, drought-affected people living in drylands in 2030 falls 13 percent with low-density tree systems and more than 50 percent with high-density tree systems (figure O.5). An important feature of tree-based systems is that, while the adoption costs must be incurred up front, the resulting benefits often take years to materialize. This can be problematic, because the long time lag to realize investment returns reduces the attractiveness of tree-based systems in the drylands, where farmers generally must focus on meeting their families' immediate consumption needs in the face of uncertain production environments. For this reason, getting farmers to adopt the technology is likely to require significant public support.

**Irrigation can provide an important buffer against droughts, particularly in the less arid parts of the drylands.** Analysis carried out for this book suggests that irrigation development is technically feasible and financially viable on 5–9 million hectares in the drylands (the number varies depending on assumptions made about capital investment costs and the minimum required level of financial returns). The area suitable for irrigation is disproportionately located

**Figure O.5** Number of drought-affected households that could be made resilient by adopting different agricultural technologies (millions)

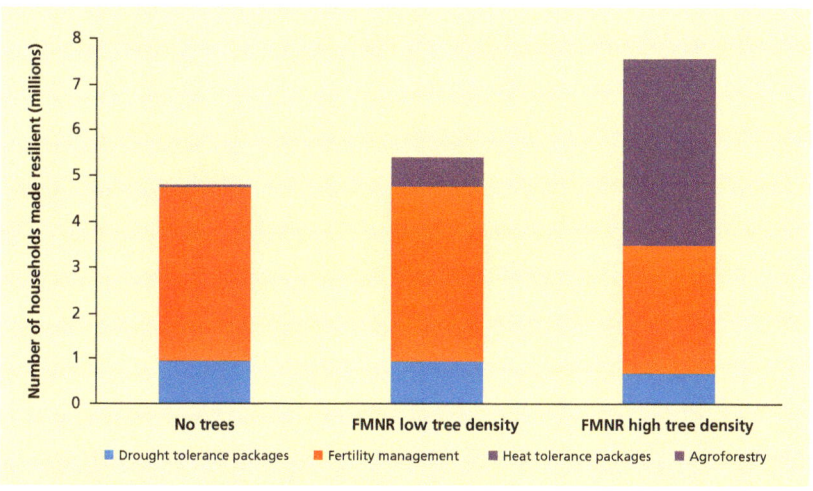

*Source:* Calculations based on World Bank data.

*Note:* FMNR = farmer-managed natural regeneration. The numbers represent households that by 2030 could become resilient to droughts by adopting different packages of resilience interventions. The figure presents aggregated results for Ethiopia, Kenya, Uganda, Senegal, Nigeria, Mauritania, Chad, Mali, Burkina Faso, and Niger.

in more humid parts of the drylands (map O.3). Up to 10 percent of the area currently being cropped in dry subhumid zones could be developed for irrigation, compared to only 2–3 percent of the area currently being cropped in arid and semi-arid zones. If this potential can be exploited, crop production losses suffered during droughts would be reduced, thereby reducing the number of drought-affected people by around 1 million, which is a 19 percent improvement compared to a package of interventions without irrigation. Most irrigation systems cannot provide reliable protection in the face of severe drought events, however. Some large-scale irrigation systems (estimated to be viable in 1.0–2.5 million hectares of dryland zones) have greater capacity to withstand more severe drought, but expansion of large-scale irrigation is likely to be constrained by extremely high capital investment costs.

## Cross-cutting interventions to enhance resilience

Other interventions discussed in this book offer additional opportunities to increase the resilience of dryland populations, as follows.

**Integrated landscape management could help to restore degraded areas in the drylands, boost productivity, and improve livelihoods.** Restoring degraded drylands by addressing the drivers of land degradation, discouraging unsustainable uses of natural resources, and scaling up improved land and water management practices can enhance the resilience of many poor and vulnerable

**Map O.3** Potential for development of small- and large-scale irrigation in Sub-Saharan Africa

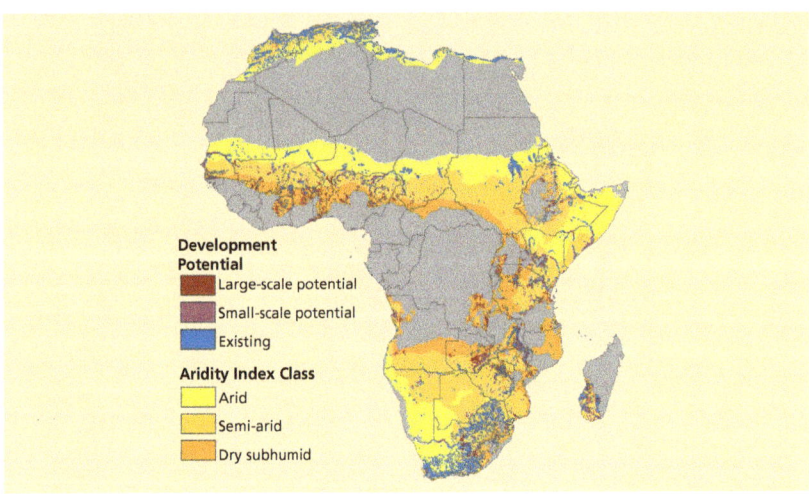

*Source:* © IFPRI. Reproduced with permission, from Xie et al. 2015; further permission required for reuse.

herders and farmers. Integrated landscape management approaches provide a potentially useful instrument for pursuing multiple objectives in the presence of a diverse set of actors. Investment in integrated landscape management programs, which support coordination and long-term collaboration among different groups of land managers and stakeholders, can enhance and safeguard restoration efforts, lower risks related to water shortages and land degradation, diversify income sources, support sustainable intensification, and reduce conflicts. Implementation of landscape approaches can be challenging, however, because of limited knowledge of the potential benefits, as well as institutional and coordination barriers to implementation.

**Reducing barriers to trade could contribute significantly to the resilience of people living in drylands by making food more available and more affordable, including after a shock hits.** The potential to develop well-integrated and competitive regional markets in African drylands is today being thwarted by barriers to trade. African agriculture continues to underperform relative to agriculture in other developing regions. While the causes of this underperformance are complex and varied, one contributing factor is the very low use of improved production inputs, especially modern plant varieties, fertilizer, crop chemicals, and animal health products. The low use of production inputs is due in part to their high cost and limited availability, a situation exacerbated by direct and indirect trade barriers. In addition to limiting the availability of vitally needed production inputs, trade barriers found throughout the drylands hamper flows of food and amplify price spikes, which can have severe implications when an extreme weather event, animal disease epidemic, or outbreak of conflict has restricted local food supplies, requiring imports of food to meet temporary shortfalls. Uncertainty caused by ad hoc trade measures also discourages investments in storage and trade infrastructure that would buffer price shocks. Initiatives to reduce barriers to trade in agricultural inputs and food will have to overcome political resistance, however, as well as pervasive mistrust between government officials and trade communities. More transparent and better information for civil society on the presence and effects of trade barriers and for government on the realities in local food markets may facilitate reforms.

## Strengthening social protection programs

Social protection programs will be a key component of successful integrated resilience strategies in the drylands, in which these programs can play two very different but complementary roles, as follows.

**Social protection programs can provide crucial safety nets to protect the most vulnerable people in times of crisis, at lower cost than humanitarian assistance.** Currently, humanitarian assistance is often the default response to droughts and other shocks. Humanitarian assistance can save lives after a shock has occurred, but it does little to strengthen resilience to future shocks. A growing body of evidence suggests that when a shock has occurred and assistance is urgently needed, it is much more cost-effective to scale up existing social protection programs, as opposed to relying on emergency aid raised through appeals. Policy makers therefore need to devise strategies for establishing and maintaining adequate safety net programs, which will mean addressing large institutional and financial challenges that many African countries presently are unable to meet.

**The ability of social protection programs to provide safety nets to all vulnerable people in drylands in times of need will come under increasing strain as a result of population growth.** Assuming that GDP continues to grow at historical rates and that future growth reduces poverty at historical rates, by 2030 the cost of providing cash transfer support to drought-affected populations is likely to be unaffordable in many dryland countries (figure O.6).

**In addition to serving as instruments that can be used to deliver safety net support, social protection programs can help build resilience at the household and community levels.** Well-designed social protection programs can facilitate the delivery of many of the best-bet interventions described above. Transfers of cash, food, or other goods offered to households in the aftermath of a drought or other shock can be accompanied by training in the use of productivity-enhancing technologies that allow vulnerable households to generate additional income. By using this additional income to build assets, these households can improve their ability to cope when the next shock hits, reducing the financing needed in future years to support shock-affected people.

**Scalable safety nets can provide cost-effective protection against many shocks, but even the strongest safety nets are unlikely to offer complete protection against some low-frequency, high-severity events.** For this reason, there will always be a need for risk transfer mechanisms, to ensure that additional fiscal resources can be mobilized at short notice to deal with the effects of severe shocks. Generally speaking, however, humanitarian assistance should be the option of last resort, rather than the alternative of choice for crisis situations.

## Enhancing preparedness with disaster risk management instruments

Disaster risk management (DRM) instruments can be key components of strategies to reduce vulnerability and increase resilience in drylands. DRM approaches can be effective in reducing sensitivity to droughts and other shocks (e.g., by putting in place screening tools and early warning systems, prioritizing infrastructure investments to increase resilience to climate shocks, or

**Figure O.6** Share of 2030 GDP required to bring the drought-affected population to the poverty line (%)

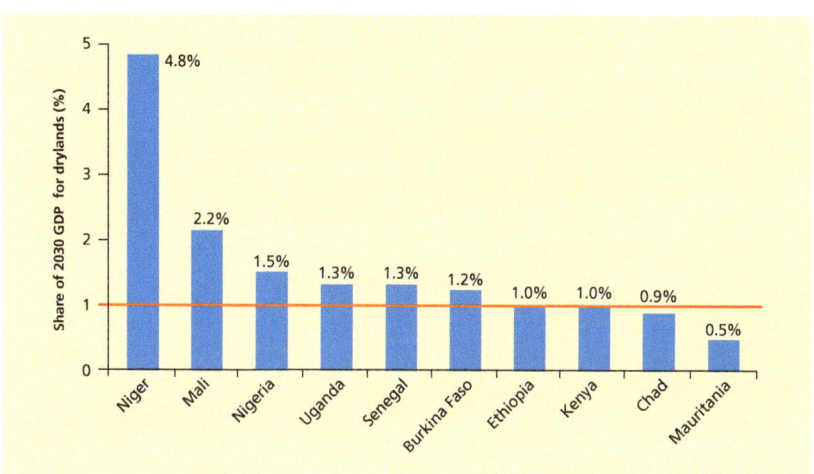

*Source:* Calculations based on World Bank data.

*Note:* The chart shows the cost of bringing, in an average year, all drought-affected people to the international poverty line through cash transfers, assuming perfect targeting (the cost is expressed as a percentage of 2030 GDP for drylands, assumed proportional to the share of the population living in drylands). The cost is calculated taking into account the country-specific depth of poverty, as proxied by 2010 poverty gap index obtained from the World Bank PovCalnet database. Figures for 2030 GDP are based on the reference growth scenario as defined in the appendix. The reference line (1 percent of GDP), indicates the consensus value in the social protection literature on the resources governments should be willing to spend in social safety nets.

introducing building codes and guidelines) as well as in improving coping capacity after a shock has hit (e.g., by supporting investments in preparedness, mobilizing sovereign disaster risk financing, making available agricultural insurance for farmers and herders, and supporting social protection programs for the poorest). DRM programs currently have limited coverage in the drylands, however, and because few programs have the capacity to scale up rapidly in response to shocks, during times of crisis most governments rely on humanitarian appeals. This is inefficient and expensive. DRM programs need to be designed and implemented in a way that is responsive to the particular dynamics of poverty and vulnerability in the drylands.

## Evaluating options: Assessing the relative merits of resilience-enhancing interventions

The scope for enhancing the resilience of dryland populations in the face of droughts was assessed using results of the umbrella model.

First, the umbrella model was used to project the likely future incidence of vulnerability in the drylands under a set of plausible assumptions about population increases, economic growth, and income distribution. By 2030, the number of vulnerable, drought-affected people living in drylands is projected to be 60 percent higher than in 2010. After 2030, the impacts of droughts and other shocks will likely become even greater as climate change increases the frequency and severity of droughts and other extreme weather events.

Next, the umbrella model was used to estimate the ability of various resilience-enhancing interventions to reduce the number of drought-affected people projected to be living in drylands in 2030. The interventions tested include: (1) improved productivity of livestock systems, (2) measures to expand the coverage and improve the productivity of irrigated agriculture, (3) measures to improve the productivity of rainfed cropping systems, and (4) improved natural resource management (in particular the use of tree-based systems).

## Potential impacts of livelihood interventions

**Interventions designed to strengthen current livelihoods could considerably reduce the number of drought-affected people living in drylands in 2030.** Adoption of the resilience-enhancing interventions would limit the increase in the number of drought-affected people to 27 percent above 2010 levels (figure O.7). This represents a significant improvement over the no-intervention BAU scenario, in which the number of drought-affected people increases by close to 70 percent compared to 2010. This result points to the importance of stepping up actions to encourage the adoption of the best-bet interventions. One needed action is to mobilize resources to pay for the effective dissemination of the technologies, estimated to range between US$0.4 and US$1.3 billion per year (depending on the assumption made about the accuracy of spatial targeting). Another needed action is to make sure that equity considerations are considered adequately in the design and implementation of resilience interventions (see box O.2).

**In some countries, improving current livelihood strategies will not be enough.** While resilience-enhancing interventions can help to *slow the increase* in the number of drought-affected people everywhere, only in some countries (Ethiopia, Uganda, and to a lesser extent Nigeria and Kenya) would the interventions *reduce the number* of drought-affected people relative to the 2010 baseline. In several countries (including Niger, Mali, Senegal, Mauritania, and to a lesser extent Chad), even after adoption of the resilience-enhancing interventions, the number of drought-affected people would increase relative to the 2010 baseline, although less than in the BAU scenario.

**Figure O.7** Potential of best-bet interventions to reduce the numbers of drought-affected people living in drylands in 2030 (2010=100)

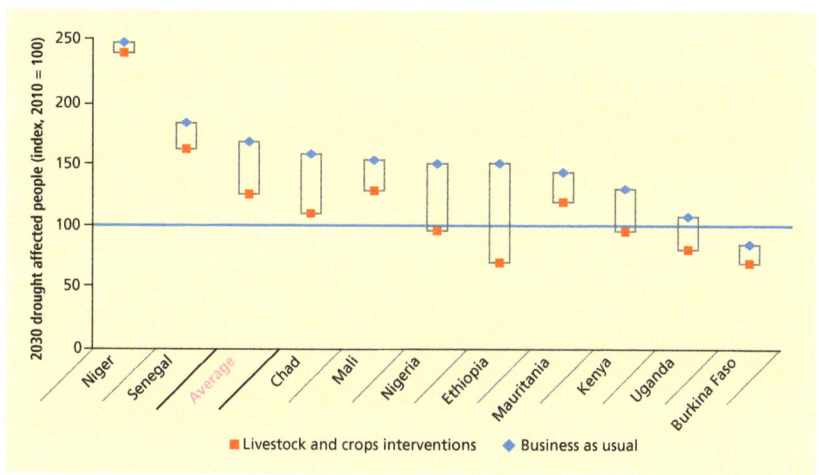

*Source:* Calculations based on World Bank data.

*Note:* The figures in the chart represent the number of dryland people projected to be dependent on agriculture in 2030 in relation to the corresponding figure in 2010. For example, a figure of 140 indicates a 40 percent increase over the 2010 level of agricultural employment. For each country, the range is defined by different scenarios of per capita GDP growth, which is expected to generate some exit of employment out of agriculture as a result of structural transformation of the economy. The details of the calculation are provided in the appendix.

**In countries where the impact of resilience-enhancing interventions is likely to be modest, fiscal realities may limit the use of social safety nets.** In countries likely to experience large increases in the number of drought-affected people, fiscal realities may limit the use of safety net programs to provide support following severe shocks. For example in Niger, Mali, and Senegal, even assuming all the resilience-enhancing interventions are adopted, the cost of using cash transfers to bring all drought-affected people up to the poverty line is likely to far exceed 1 percent of GDP, the consensus value in the social protection literature on the resources governments should be willing to spend on social safety nets (figure O.8). In these countries, the policy choices boil down to reducing the number of people covered by social safety nets, limiting the amount of support provided per person, or relying on humanitarian assistance to fill the fiscal gap.

## The fiscal dividend of resilience-enhancing interventions: A country typology

In considering the potential of the best-bet interventions to reduce vulnerability and increase resilience among populations living in drylands, it is important to

## BOX 0.2

### Recognizing equity considerations

Cost-effectiveness is one factor that policy makers and development practitioners must take into account in designing dryland development policies and programs, but it is not the only factor. As everywhere, efforts to reduce vulnerability and build resilience in drylands are complicated by political economy factors. Because any change in the status quo is likely to bring benefits to some groups and impose costs on other groups, the desirability of alternative interventions must always be assessed taking into account equity considerations.

For example, expanding irrigation schemes into previously uncultivated land benefits the farmers who gain access to irrigation services, but it harms pastoralists who had been able to take advantage of feed resources on the previously uncultivated land. Conversely, improving veterinary services to reduce animal mortality rates benefits the livestock keepers who see their herds increase, but it harms the farmers who subsequently experience more frequent invasions of their fields by free-roaming animals.

Development interventions are often portrayed as activities that can improve the welfare of all, but since interventions inevitably play out against established distributions of wealth and power, they are rarely Pareto efficient—usually there are winners and losers. These considerations loom especially large in many dryland regions of Africa, where competition for scarce resources in a context of political instability has fueled recurring cycles of conflict.

note that the interventions will have two types of effects—direct and indirect. Investments in livestock and crop farming systems will directly reduce the number of drought-affected people by improving the productivity and sustainability of current livelihood strategies. In addition, these investments will indirectly contribute to improved resilience in the drylands by freeing up public resources that would otherwise have to be used for emergency responses. These resources can be redirected to programs designed to strengthen the resilience of vulnerable segments of the population. They can be thought of as the "fiscal dividend" produced by resilience-enhancing interventions.

The presence or absence of this fiscal dividend can be used to define a policy-relevant typology of countries, distinguished according to the differing ability of the resilience-enhancing interventions to reduce the cost of protecting vulnerable livelihoods in the drylands.

In Niger, Mali, and Senegal (referred to here as Group A), where opportunities to reduce sensitivity and increase coping capacity among vulnerable households are limited, the resilience-enhancing interventions have the potential to reduce the cost of supporting drought-affected people using safety nets, but the

**Figure O.8** Cost of cash transfers needed to support drought-affected people in drylands in 2030, with and without interventions (% of GDP)

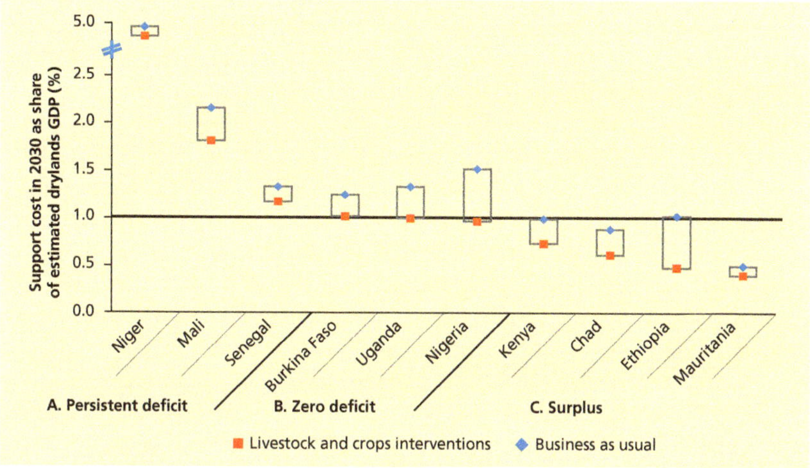

Source: Calculations based on World Bank data.

Note: The vertical axis has been trimmed to avoid the distorting effect of the outlier (Niger). The chart shows the cost in an average year of bringing all drought-affected people to the international poverty line through cash transfers, assuming perfect targeting (the cost is expressed as a percentage of 2030 GDP for drylands, assumed proportional to the share of the population living in drylands). The cost is calculated taking into account the country-specific depth of poverty, as proxied by the 2010 poverty gap index obtained from the World Bank PovCalnet database. Figures for 2030 GDP are based on the reference growth scenario as defined in the appendix. For each country, the higher end of the range is the business as usual scenario; the lower end of the range is a scenario of adoption of the productivity-enhancing technologies analyzed throughout the book. The difference between the higher and lower end of the range is the benefit in terms of savings of the cash transfers required to bring all drought-affected people to the poverty line. The reference line (1 percent of GDP) indicates the consensus value in the social protection literature on the resources governments should be willing to spend in social safety nets.

residual cost remains well above the 1 percent of GDP benchmark. Many people living in drylands in these countries are likely to remain vulnerable, even after the resilience-enhancing interventions have been implemented and safety net programs put in place. In these countries, where coping capacity is likely to remain limited and sensitivity to shocks high, an important policy priority is to reduce overall exposure by way of interventions to promote alternative livelihoods both inside and outside of drylands.

In Burkina Faso, Uganda, and Nigeria (referred to here as Group B), where opportunities to reduce sensitivity among vulnerable households are somewhat greater, the resilience-enhancing interventions combined with safety net spending at the 1 percent of GDP level fully cover the drought-affected population living in drylands. But after the resilience-enhancing technologies have been disseminated and safety net programs strengthened, few resources would be left that could be invested in helping drought-affected people become resilient over the longer term. In Group B countries, the need to promote alternative livelihood strategies is likely to be less urgent than in Group A countries, but these countries will have little or no fiscal space to respond to contingencies (e.g., extreme drought events), and, more importantly, they will have limited resources available to invest in making vulnerable populations more resilient over the longer term. An important priority for these countries is to develop mechanisms for rapidly mobilizing contingent financing to respond to occasional extreme crises.

In Kenya, Chad, Ethiopia, and Mauritania (referred to here as Group C), where opportunities to reduce sensitivity and increase coping capacity among vulnerable households are considerable, once the resilience-enhancing interventions have been implemented, all remaining drought-affected people living in drylands can be supported by safety nets, at a combined cost well below 1 percent of GDP. In these countries, resources that previously might have been needed to respond to droughts and other shocks can in future be invested in making dryland populations more resilient over the longer term. Key priorities for Group C countries include scaling up investments in resilience-enhancing interventions (to turn into reality the potential fiscal dividend) and identifying strategies for productively investing the fiscal dividend.

## Promoting new livelihoods to manage the transformation

The results of the umbrella modeling exercise highlight the possibilities and the limitations of interventions designed to improve the productivity of current livelihood strategies in the drylands. In considering the policy implications, however, it is important not to lose sight of the fact that the future will not be identical to the past.

**Rapid population growth in drylands will exacerbate many existing challenges, but population growth will also bring new opportunities.** Increased population density in the drylands will create opportunities for profitable commerce and trade, increased economic specialization, and enhanced value addition. Similarly, increased population density in the drylands will generate economies of scale in the provision of essential public services (such as education, health care, water and sanitation, communications, and security), thereby reducing the corresponding cost. In short, population growth in the drylands could prove vital in overcoming the traditional problem that has contributed to the underdevelopment of many dryland areas—namely, that the sparse population distributed over vast areas has made markets thin and costly, discouraging both public and private investment in the provision of goods and services.

**Seizing the emerging opportunities will be possible only to the extent that higher population density combined with increasing expropriation by state**

and external investors will not lead to increased competition for natural resources, especially land, water, and biomass. Increased competition will likely put added pressure on resources, which could give rise to increased conflict. For this reason, as population growth outstrips the ability of current livelihood strategies to provide adequate incomes for all, public policy will have to focus on generating new livelihoods, less reliant on natural capital, and more on human and physical capital.

Livestock-keeping and crop farming can continue to be important components of the livelihood strategies of people living in drylands. These activities will have to be complemented by new sources of income, however—not only post-harvest value-adding activities related to the processing of agricultural products, but also employment in the services and manufacturing sectors. Because this change will require exits from livelihoods based on agriculture and natural resources and migration to employment in other sectors, the solution to the problems of drylands to a significant extent will come from outside the drylands.

## Policy recommendations

Enhancing the resilience of people living in the drylands will require a combination of interventions to improve current livelihoods and interventions to strengthen safety nets. An overarching recommendation emerging from the analysis reported in this book is that policy makers in dryland countries and their partners in the development community may want to look more closely at each of the two types of interventions, to assess their potential in more detail than has been possible here, taking into account local circumstances and development priorities. The Country Programming Framework prepared in the aftermath of the 2011 drought by the countries of the Horn of Africa is an important step in that direction. Strategic plans formulated at the country level and at the regional level should be updated regularly and broadened and deepened as new knowledge becomes available, focusing especially on the medium to long term and quantifying to the extent possible the technical and financial potential of alternative interventions. With respect to the two types of interventions, this book presents detailed recommendations, which are summarized in box O.3.

Improving current livelihood activities and strengthening social protection programs have significant potential to reduce vulnerability and enhance resilience of populations living in drylands, but both strategies are likely to face limits. The scenario analysis carried out using the umbrella model shows that even if current livelihood strategies can be improved and social protection programs strengthened, significant numbers of households will remain vulnerable

**BOX O.3**

## Summary of recommendations to make current livelihoods more resilient

(1) Livestock
- Increase production of meat, milk, and hides in drylands by developing sustainable delivery systems for animal health, promoting increased market integration, and exploiting complementarities between drylands and higher rainfall areas.
- Enhance the mobility of herds by expanding and ensuring adequate and equitable year-round access to grazing and water and by improving security in pastoral zones.
- Develop livestock early warning systems (LEWSs) and early response systems to reduce the adverse impacts of shocks.
- Identify additional and alternative livelihood strategies, including through systems of payment for environmental services.

(2) Farming
- Accelerate the rate of varietal turnover and increase availability of hybrids.
- Improve soil fertility management.
- Improve agricultural water management.
- Promote the development of irrigation, including both rehabilitation of existing capacity, and expansion, up to the viable potential (a maximum of about 10 million more hectares); and focusing on small-scale systems, with good access to markets for cash crops.

(3) Natural resource management
- Promote farmer-managed natural regeneration (FMNR) to establish a range of beneficial trees throughout the drylands.
- Invest in tree germplasm multiplication and promote planting of location-appropriate high-value species, especially in dry subhumid areas.
- Develop opportunities to add value to tree products produced in the drylands.

(4) Social protection
- Establish and gradually expand the coverage of national adaptive safety net programs that promote resilience of the poorest people.
- Use social protection programs to build capacity of vulnerable households to climb out of poverty, but maintain the ability to provide humanitarian assistance in the short run.
- Respond to emergencies by scaling up existing programs, rather than relying on appeals for humanitarian assistance.
- Tailor social protection programs to address the unique circumstances of dryland populations.

to droughts and other shocks while lacking the resources to cope effectively when a drought strikes. For these households, policy makers will need to devise strategies to facilitate the transition to alternative livelihood activities. While the results of the umbrella modeling exercise help in defining the extent to which alternative livelihood strategies will be needed, this book does not present detailed analysis of the policy reforms and the complementary investments in human and physical capital that will be needed to help poor and vulnerable households in the drylands transition out of natural resource–based livelihoods to productive employment in other sectors, nor does it make specific recommendations relating to these policy reforms and investments. These types of interventions fall outside the scope of the present inquiry, and further work will be needed to cover them adequately.

## Note

1. First proposed by Budyko (1958) and subsequently endorsed by the United Nations Environment Programme as part of the preparations for the United Nations Conference on Desertification.

## References

Budyko, M. I. 1958. *The Heat Balance of the Earth's Surface.* Translated by N. A. Stepanova. Washington, DC: U.S. Dept. of Commerce.

De Haan, C., E. Dubern, B. Garancher, and C. Quintero. 2014. *Pastoralism Development in the Sahel: A Road to Stability?* Nairobi: World Bank Global Center on Conflict, Security, and Development.

OECD (Organisation for Economic Co-operation and Development). 2015. International Development Statistics (IDS) online databases. OECD, Geneva. http://www.oecd.org/dac/stats/idsonline.htm.

Scandizzo, P.L., S. Savastano, F. Alfani, and A. Paolantonio. 2014. "Household Resilience and Participation in Markets: Evidence from Ethiopia Panel Data." Unpublished document, World Bank, Washington, DC.

Xie, Hua, Weston Anderson, Nikos Perez, Claudia Ringler, Liang You, and Nicola Cenacchi. 2015. "Agricultural Water Management for the African Drylands South of the Sahara." Background report for the Africa Drylands Study. International Food Policy Research Institute, Washington, DC.

# Part A

# Key Issues and Challenges

## Chapter 1

# The Central Role of Drylands in Africa's Development Challenge

*Michael Morris, Raffaello Cervigni, Zhe Guo, Jawoo Koo*

The dramatic humanitarian crises caused by the crippling droughts that have ravaged the Horn of Africa and the Sahel in recent years have once again brought to the forefront of the development debate the chronic vulnerability of many of the people living in dryland regions of Sub-Saharan Africa. Breaking the recurring cycle of drought, suffering, and impoverishment will not be easy. To design the resilience-enhancing interventions needed to shield people living in drylands from the droughts and other shocks that they regularly experience, policy makers and donor partners must be able to identify the vulnerabilities that keep so many households mired in poverty, project how these vulnerabilities will evolve over time, and evaluate the relative advantages and disadvantages of interventions that have the potential to improve and stabilize the livelihood strategies on which the most vulnerable households depend.

## Definition of "drylands"

What exactly are "drylands"? While commonly used, the term has different interpretations. For reasons of simplicity, and consistent with widespread practice, in this book "drylands" are defined on the basis of the Aridity Index (AI). Under this approach, which has been endorsed by the 195 parties to the United Nations Convention to Combat Desertification (UNCCD) and which is also being used by the United Nations Food and Agriculture Organization (FAO), drylands are defined as regions having an AI of 0.65 or less (for details, see UNEP 1997). Drylands furthermore can be sub-divided into four zones: hyper-arid (AI 0–0.05), arid (AI 0.05–0.20), semi-arid (AI 0.20–0.50), and dry subhumid (AI 0.50–0.65).

Because the hyper-arid zone is incapable of supporting crop and livestock production activities, it is very sparsely populated, making it of little interest to policy makers. For purposes of this book, "drylands" is therefore defined as the area characterized by an AI of 0.05–0.65, encompassing the arid, semi-arid, and dry subhumid zones (map 1.1).

## Reasons for concern about drylands

Defined based on the AI, as above, drylands in Sub-Saharan Africa cover about 13.9 million square kilometers (km²) (map 1.1). They are home to about 425 million people and account for 70 percent of the region's cropland, 66 percent of cereal production, and 82 percent of livestock holdings (figures refer to 2010). Most drylands are marginal environments characterized by challenging agroclimatic conditions and endowed with limited resources to support primary production activities, such as livestock-keeping and farming, so they tend to be hotspots of natural resource degradation. In addition, because of the remoteness of many drylands, the rule of law is often weak, leading to unusually high levels of conflict in drylands that further exacerbate the vulnerability of local populations. The fragility of current livelihood strategies in drylands is often compounded by the social and political marginalization of many of the groups

**Map 1.1** Dryland regions of Sub-Saharan Africa, defined in terms of the Aridity Index

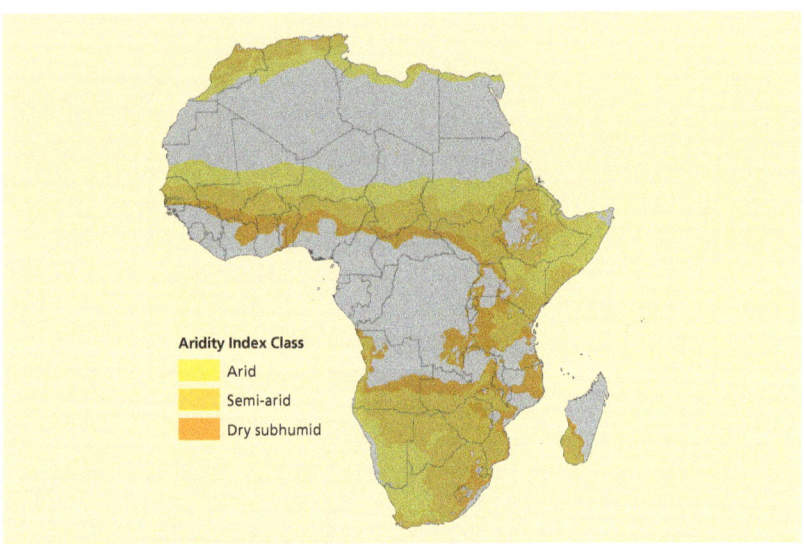

*Source:* ©Harvest Choice, IFPRI, 2015. Reproduced, with permission from Zhe Guo, 2015; further permission required for reuse.

that live in drylands, which muffles their voices and limits their ability to influence political processes that affect their well-being (Kerven and Behnke 2014). For all of these reasons, drylands are home to a large share of the region's poor, as well as many of those lacking access to basic services, such as health care, education, water, and sanitation.

Today in the drylands, frequent and severe shocks—especially those caused by recurring extreme and prolonged droughts—limit the livelihood opportunities available to millions of poor households and undermine efforts to eradicate poverty. In the absence of robust social protection systems and rapidly scalable safety nets, these shocks cause large drains on government budgets and consume a significant portion of the region's international development assistance. As a result, scarce resources are diverted away from pursuing longer-term development goals and redirected to mobilizing costly short-term responses to humanitarian crises. In 2011 around US$4 billion was spent on humanitarian assistance to the Sahel and Horn of Africa alone, equivalent to 10 percent of total Overseas Development Assistance to all of Sub-Saharan Africa (OECD 2015).

If the current situation is precarious, the future promises to be even more challenging. By 2030 the population living in drylands is expected to grow by 58–74 percent (depending on the fertility scenario), putting increased pressure on a resource base already severely stretched. Over the same period, climate change could result in an expansion of the area classified as drylands (up to 20 percent under some scenarios), bringing more people into environments in which livelihood options are limited and in which opportunities to ensure resilience are severely constrained. Higher population density in the drylands will put additional pressure on a fragile resource base, pushing it in some cases beyond its natural regenerative capacity. This could escalate social conflicts over land, water, and biomass. At the same time, higher population density will bring new development opportunities linked to greater market size, increased economic specialization, and enhanced value addition, as well as possibilities to achieve cost savings in the provision of vital services such as education, health care, water and sanitation, energy, communications, and security.

Because the ongoing transformation of the drylands is being propelled by demographic drivers that have a great deal of momentum, it is in many respects inevitable. In this context the key question for policy makers is how best to manage the coming demographic, social, and economic changes to achieve the best possible outcomes. As governments and donor partners contemplate the design of the next generation of policies and programs for the drylands, it is important to know whether traditional pursuits, especially livestock-keeping and crop farming, can be made sufficiently productive and stable in the face of demographic, economic, and climatic change to provide secure livelihoods for the entire population. If the scope for sustainable intensification is limited, fundamental transformations of the predominant livelihood systems may be needed to avoid increasingly frequent and ever more consequential humanitarian crises.

The stakes extend far beyond the drylands themselves. The facts that drylands are home to such a large share of Africa's population and account for such a large proportion of the region's total food supply mean that population dynamics and agricultural activities in drylands affect the demographics and food security of the continent as a whole. In addition, because many people living in drylands lack the resilience needed to recover from droughts and other shocks, drylands are home to disproportionate numbers of the region's poor. For this reason, it will be impossible to meet many of the long-term development goals shared by African governments and donor partners—including the World Bank Group's twin goals of reduced poverty and shared prosperity—unless the problems of drylands are addressed.

## Objectives of this book

What are the prospects for making poor households living in dryland regions of Africa resilient in the face of the crippling droughts and other shocks that so regularly disrupt their livelihood activities, often with devastating consequences? Will economic growth alone solve the problem by providing these households with the resources needed to protect themselves from the effects of droughts and other shocks? To what extent can technical interventions increase the productivity, stability, and sustainability of the livestock-keeping and crop production activities on which most of these households depend? If economic growth and technical interventions are likely to fall short, what other options are available to secure the well-being of vulnerable populations?

The World Bank Group, in collaboration with many partners—including FAO, IFPRI (International Food Policy Research Institute), ILRI (International Livestock Research Institute), ICRAF (World Agroforestry Centre [known before 2002 as the International Centre for Research in Agroforestry, ICRAF]), ICARDA (International Center for Agricultural Research in the Dry Areas), ICRISAT (International Crops Research Institute for the Semi-Arid Tropics), CIRAD (Agricultural Research for Development), CILSS (Permanent Interstates Committee for Drought Control in the Sahel), and WRI (World Resources Institute)—recently carried out a major study designed to address these questions. Taking advantage of the rich set of data, knowledge, and analytical tools that have become available in recent years, the study team developed an original quantitative framework that allowed it to project through 2030 patterns of vulnerability in African drylands and test the likely impacts of a series of policy reforms and technical interventions. This book presents key findings and recommendations emerging from the study. Focusing primarily on the two biggest vulnerability hotspots—the Sahel region of West Africa and the Horn of Africa region in East Africa—the book sheds light on the factors contributing to vulnerability among dryland populations, identifies strategies for enhancing the resilience of the millions of households that depend on traditional livelihood activities, such as livestock-keeping and farming, and draws a number of conclusions that have important implications for policy making and program design.

The book has three principal objectives:

1. Characterize current and future challenges to reducing vulnerability and increasing resilience in the drylands of Sub-Saharan Africa.

2. Identify interventions that can enhance the resilience of populations living in the drylands, estimate the cost of these interventions, and assess their effectiveness.

3. Provide an evidence-based framework that can be used to improve decision making on alternative options to enhance resilience.

Based on a comprehensive review of the evidence, the book argues that two distinct yet complementary approaches will be needed to reduce vulnerability and increase resilience in dryland regions of Sub-Saharan Africa, as follows:

1. **Improve current livelihood activities:** For the foreseeable future, most of the people living in drylands in East and West Africa will continue to make their living from herding and farming. For that reason it will be important to make current livelihood strategies (especially pastoralism, agro-pastoralism, and crop farming) more productive and more resilient. The book therefore examines in detail technical options for improving current livelihood strategies, and it uses a range of modeling approaches to assess the potential impacts of different technical interventions in terms of making existing livelihood strategies more productive and more resilient.

2. **Strengthen social protection programs including rapidly scalable safety nets:** In many parts of the drylands, even the most productive, stable, and sustainable livelihood activities will not be fully immune to the effects of droughts and other shocks. For this reason, it will be necessary to put in place social protection programs including rapidly scalable safety nets to address the needs of those lacking the resilience to cope effectively with the effects of droughts and other shocks. Therefore this book examines in detail the feasibility and likely cost of using safety nets and other types of social protection programs to provide assistance to those in need.

Improving current livelihood activities and strengthening social protection programs have significant potential to reduce vulnerability and enhance resilience of populations living in drylands, but both are likely to face limits, particularly in the face of technological, financial, and fiscal constraints. In light of these limits, policy makers will need to consider a third set of interventions, namely, those that encourage dryland populations to switch to alternative livelihood activities that are less vulnerable to droughts and other shocks. By assessing the scope and limitations of the first two types of interventions, this book helps define the importance, across the group of countries analyzed, of the third type of intervention. The book does not attempt to identify or analyze in detail the alternative livelihood activities that may offer the brightest prospects for dryland populations in East and West Africa; that would require high-level analysis of long-term structural transformation processes affecting the dryland countries, combined with a series of "deep-dive" analyses focusing on associated topics, such as demographics, health, education, and employment. Those tasks fall outside the scope of the present study and remain topics for future research.

## Value-added of this book

Several features of this book distinguish it from the many other books, studies, and reports that have focused on questions of vulnerability and resilience in the drylands of Sub-Saharan Africa.

First, the study whose results are presented here was carried out by a large team of collaborators representing the full range of organizations that are active in dryland development initiatives, including government agencies, regional organizations, multilateral development agencies, research institutes, and nongovernmental organizations. These many collaborators brought with them a range of perspectives and a wealth of knowledge that are reflected in a study of unparalleled scope and unprecedented depth.

Second, the study team developed a comprehensive analytical framework that incorporates insights derived from work done in many different sectors. In addition to exploring opportunities to increase productivity through sustainable intensification of current livelihood strategies (such as livestock-keeping and crop production), the analytical framework considers opportunities to reduce vulnerability and increase resilience in drylands through investments in social protection instruments, improved connectivity, and disaster risk management programs.

Third, the approach used by the study team is evidence based. Because of the technical difficulty and high cost of conducting surveys in sparsely populated and physically remote dryland areas, credible data on the activities of dryland populations are often lacking. For this reason, the study team invested considerable time and effort into assembling data sets that could be used to carry out rigorous quantitative analysis. Modeling efforts focused on a number of areas, including the dynamics of dryland livestock systems, the technical and economic potential for irrigation development in drylands, the potential for sustainable intensification of rainfed cropping systems in drylands, and the likely evolution of vulnerable populations living in drylands.

## Limitations of the book

This book presents a wealth of analytical results that go beyond previously available knowledge, but even so, it suffers from a number of shortcomings. Three are worth mentioning. First, the coverage is geographically limited. The primary focus is on the Sahel and the Horn of Africa, two hotspot regions featuring extensive dryland areas that have been particularly hard hit in recent decades.

Despite the best efforts of the study team to find and exploit all available data sets for these two focal areas, gaps remain in the empirical record, particularly in countries that have suffered extended periods of conflict and in countries that have lacked capacity to collect, process, and publish statistics. The scope of coverage is relatively good in the Sahel, where the main resilience analysis covers approximately 85 percent of the projected 2030 population. It is more limited in the Horn of Africa, where the main resilience analysis covers approximately 69 percent of the projected 2030 population.

Second, even in areas for which data are available, the data do not cover all relevant topics. There is much we still don't know about drylands—with respect to their physical features; the characteristics of the resident plant, animal, and human populations; and the dynamic processes that shape the interactions between physical features and living communities.

Third, despite the efforts of the study team to adopt a broad view in analyzing vulnerability and resilience in the drylands, because of time and resource limitations, it was necessary to restrict the focus of analysis; as a result, certain topics were not covered in depth. For example, while it is well known that conflict contributes to the vulnerability of many of the people living in the Sahel and the Horn of Africa, the topic of conflict was not covered in depth, as this would have required extensive analysis from a social, cultural, and political economy perspective of the historical forces that over time have shaped distributions of wealth, power, and influence and given rise to present-day conflicts.

## References

Kerven, C., and R. Behnke, eds. 2014. "Human, Social, Political Dimensions of Resilience." Unpublished paper, FAO, Rome.

OECD (Organisation for Economic Co-operation and Development). 2015. International Development Statistics (IDS) online databases. OECD, Geneva. http://www.oecd.org/dac/stats/idsonline.htm.

UNEP (United Nations Environment Programme). 1997. *World Atlas of Desertification.* 2nd edition. London: UNEP.

## Chapter 2

# Resilience and its Determinants: A Conceptual Framework

*Michael Morris, Raffaello Cervigni*

## Analyzing resilience: Conceptual and data challenges

Prospects for sustainable development of drylands are assessed in this book through the lens of resilience. But what exactly is meant by resilience? While there are many ways to define resilience, most definitions highlight the ability of people or ecosystems to withstand and recover from short-term shocks, in this case understood to mean mainly droughts. The approach used in this book is consistent with common practice, but it is not as comprehensive as some approaches in that it focuses more on people than on ecosystems (see box 2.1 ).

---

**BOX 2.1**

### Resilience: Ecological and socioeconomic approaches

Consistent with the focus of many governments and of much of the development community, this book uses the concept of resilience as a framework for assessing the effectiveness of potential interventions to increase incomes, reduce poverty, and improve the welfare of people living in drylands. It is important to recognize, however, that the concept of resilience is used here in a way that differs from the way it is often used in the biological and human sciences, where it has a long and useful intellectual lineage. In the biological and human sciences, resilience typically refers to systems, not individuals, and a distinction is often made between the *resilience* of a system and the *stability* of a system (Holling 1973 cited in Kerven and Behnke 2014). Resilience refers to the persistence of a system and its ability to absorb change and disturbance and maintain the same relationships. In contrast, stability represents the ability of a system to return to an equilibrium state after a temporary disturbance; the more rapidly it returns and the less it fluctuates, the more stable it is. Critically, resilience may come at a cost in terms of exposure to risk and the maintenance

*(continued next page)*

---

**Box 2.1** *(continued)*

of a diverse set of responses to risk. Resilience may also come at a cost to individual organisms; what is resilient, survives, and persists is the system or community, not an individual component within it.

In this book the primary focus is on people—communities, households, and their individual members—rather than on livelihood systems as such. The distinction is important because it allows us to recognize that even though the livelihood systems found in drylands may be resilient over the longer term, they also tend to be unstable in the short to medium term, subjecting the people who rely on those livelihood systems to significant swings in fortune when shocks hit, as they frequently do. Systems analysts argue correctly that dryland livelihood systems, such as pastoralism, agro-pastoralism, and farming, have demonstrated a remarkable ability to recover from major shocks; however, government authorities and development practitioners cannot simply ignore the considerable instability that occurs along the way. When shocks hit—for example, the severe droughts that ravage many dryland areas on a regular basis—it may be true that the prevailing livelihood systems are likely to recover eventually, but in the short run the humanitarian consequences are severe: crops fail, animals die, and people go hungry and eventually starve. Governments, the development community, and humanitarian organizations obviously cannot ignore these short-term effects.

This book uses a simple conceptual framework for analyzing resilience in African drylands, one that attempts to reconcile the key features of resilience with the constrained realities of data availability in Africa. The starting point for analysis is the observation that drylands tend to be particularly exposed to droughts, which in combination with other factors contributes to especially unfavorable development outcomes in drylands.

Using Nigeria as an example, the Palmer Drought Severity Index (PDSI) can be used to show the differential occurrence and severity of drought phenomena in dryland zones (map 2.1). Over the period 1950–2008, severe drought events lasting two or more years occurred with much greater frequency in the drier northern part of Nigeria than in the more humid central belt or the well-watered southern part. The extremely dry northwestern part of the country was a particular hotspot, with severe drought events occurring in more than 30 percent of all years.

The droughts that disproportionally affect drylands contribute to consistently negative development outcomes. Evidence of this comes from a series of surveys carried out between 2008 and 2013 in six countries (Ethiopia, Malawi, Niger, Nigeria, Tanzania, and Uganda) under the World Bank-supported Living Standards Measurement Surveys (LSMS). While these six countries do not

**Map 2.1** Distribution of drought hotspots in Nigeria, 1950–2008 (%)

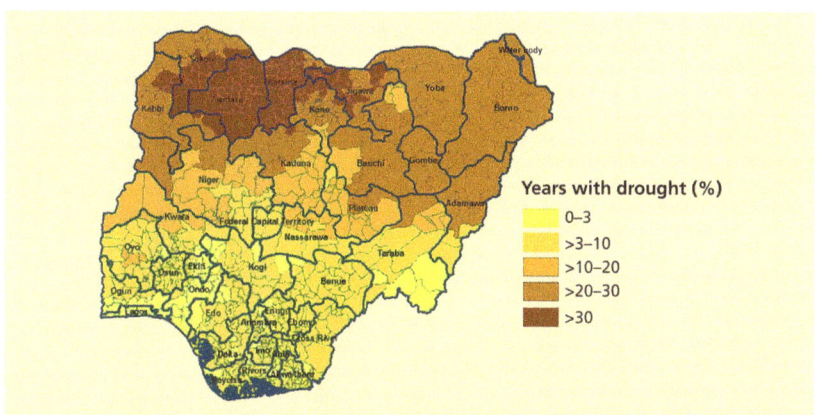

*Source:* Strzepek, Strzepek, and Neumann 2014.
*Note:* Severe drought is defined as Palmer Drought Severity Index (PDSI) less than –3.

represent the entire range of countries that are the focus of this book, all contain significant dryland areas, and as such, they provide useful insights that are of relevance to drylands generally.

Consistent with expectations, across the six countries included in the LSMS sample, the incidence of poverty is higher in dryland zones than in other more humid zones, and the poverty headcount increases with the level of aridity (figure 2.1). The overall averages mask considerable variability between individual countries, especially in terms of the level of poverty, but the relative incidence

**Figure 2.1** Poverty headcount by aridity zone, selected countries, 2010 (%)

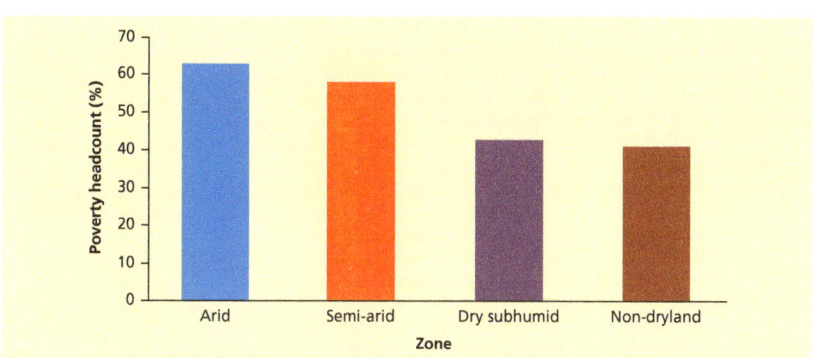

*Source:* D'Errico and Zezza 2015.
*Note:* Based on data collected in selected countries with significant drylands: Ethiopia, Malawi, Niger, Nigeria, Tanzania, Uganda.

across aridity zones is quite consistent, with the poverty headcount being higher in more arid zones in all countries except Tanzania.

Not surprisingly, the higher levels of poverty observed in dryland zones are associated with higher levels of food insecurity (figure 2.2), which in turn affects health indicators (figure 2.3).

**Figure 2.2** Average food consumption score, drylands vs. non-drylands, 2010

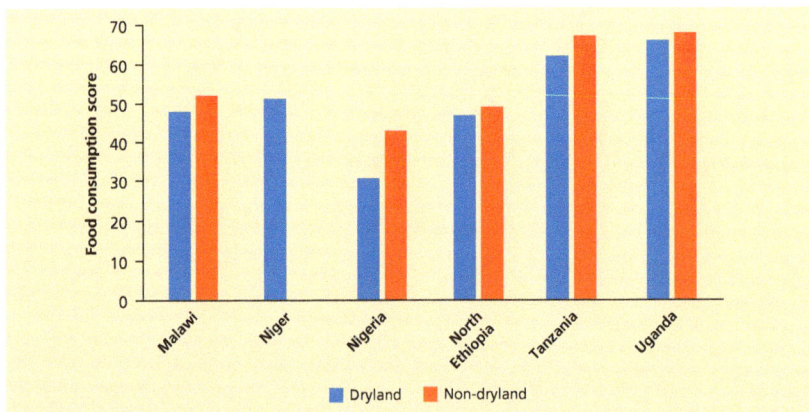

*Source:* D'Errico and Zezza 2015.

*Note:* Food security is a complex phenomenon that cannot be easily captured by any one indicator. In this figure, the Food Consumption Score (FCS, Wiesmann et al. 2009) is used to approximate food security. The FCS is based on the weighted frequency (number of days in a week eaten) of 8 food groups: staples, pulses, vegetables, fruits, meat/fish/egg, milk, sugar, oil. A higher score is purported to indicate a greater food security.

**Figure 2.3** Underweight children, drylands vs. non-drylands, 2010 (%)

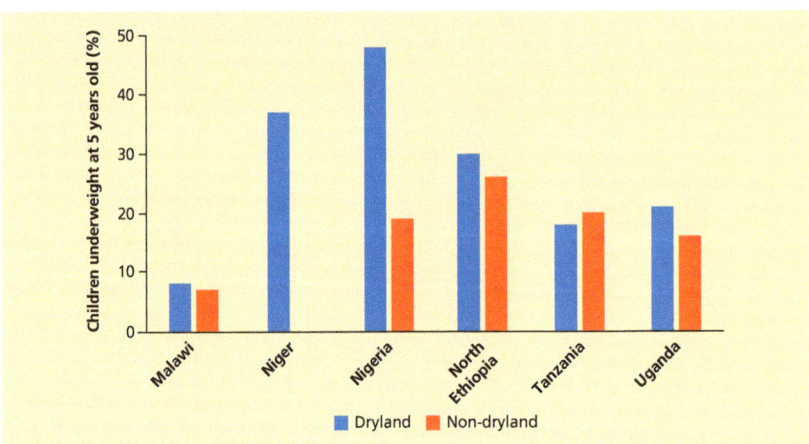

*Source:* D'Errico and Zezza 2015.

The LSMS data show a link between drylands and negative development outcomes, but the picture is static, because the LSMS data were collected in each country through a one-off survey. Resilience refers to the ability of households to cope successfully with droughts and other shocks; since coping activities take place over time, resilience is an inherently dynamic concept. For this reason, accurate measurement of resilience requires panel data—that is, data collected from households at multiple points in time. In addition, because resilience is complex and multi-faceted, it cannot be understood unless data are available on the following types of variables:

1. The frequency and severity of droughts or other shocks
2. Multiple household characteristics that determine why some households respond better than others to shocks of similar nature
3. Development outcomes (e.g., poverty, nutrition score, health status)

In many countries in East and West Africa, high-quality data on these variables are not available. In the relatively few cases where high-quality data are available, frequently they are available for only one point in time, which is severely limiting when it comes to analyzing resilience, because data from a single point in time reveal little about the movements by individual households in and out of poverty. This is an important limitation, as there may be significant differences in the causes of—and eventual solutions to—chronic structural poverty on the one hand and transient stochastic poverty on the other (Barrett and Carter 2013; Carter and Barrett 2006).

Scandizzo et al. (2014) analyzed the dynamics of vulnerability and resilience at the household level using a unique set of panel data for Ethiopia collected over a 16-year period (four rounds of surveys were conducted between 1994 and 2009). These authors found that over the 16 years covered by the panel, many households in the sample transitioned in and out of poverty, so during a period when the overall poverty headcount was gradually coming down, the fortunes of individual households tended to be highly variable (table 2.1).

**Table 2.1** Shares of households in transition across poverty status in Ethiopia, 1994–2009 (%)

| Year | Move into poverty (%) | Stay poor (%) | Stay non-poor (%) | Move out of poverty (%) |
|------|-----------------------|---------------|-------------------|-------------------------|
| 1994 | n.a. | n.a. | n.a. | n.a. |
| 1999 | 18 | 17 | 45 | 20 |
| 2004 | 19 | 16 | 48 | 16 |
| 2009 | 18 | 17 | 46 | 19 |

*Source:* Scandizzo et al. 2014.

The amount of movement among households in the sample is surprisingly high. On average in any given year, approximately 18–19 percent of households started out non-poor and fell into poverty, 16–17 percent of households started out poor and stayed poor, 16–20 percent of households started out poor and climbed out of poverty, and 45–48 percent of households started out non-poor and remained non-poor. An important insight emerging from these results is that conventional measures of poverty can conceal as much as they reveal. Within a given population, many combinations of completely different lifetime stories can give rise to the same poverty headcount. This means that policy makers and development practitioners need to have a good understanding of the dynamic factors that determine vulnerability and resilience at the household level if they are going to design effective dryland development policies.

In the absence of panel data, efforts to understand the determinants of vulnerability and resilience in drylands typically rely on cross-sectional data. Analysis of cross-sectional data can produce important insights into the factors associated with negative development outcomes, but these insights often lack the degree of specificity needed for the design of policies specific to drylands (see box 2.2).

## The determinants of resilience

Mindful of the limitations of currently available socioeconomic and climatic data, which make it difficult to estimate directly the resilience of households living in the drylands, this book uses an approach based on the identification of the likely *determinants* of resilience, rather than on the measurement of resilience itself.

Resilience—understood here to mean the ability of individuals, households, and/or communities to withstand and respond to droughts and other shocks[1]— is determined by three factors:

- **Exposure** can be defined as the nature and degree to which the income-generating assets of a household are located in places where they are subject to droughts and other shocks. A household whose assets are located in an area in which severe droughts occur once in every 5 years on average is more exposed than a household whose assets are located in an area in which severe droughts occur once in every 15 years on average. Exposure is an exogenous dimension of vulnerability, that is, it is beyond the control of the household in the short run.

- **Sensitivity** is the degree to which a household is affected by droughts and other shocks. For a given level of exposure, a household that derives a large share of its income from drought-affected activities (e.g., rainfed cropping, pasture-based livestock production) will have a higher sensitivity to

## The challenge of analyzing dryland poverty through cross-country analysis

In an effort to identify the main correlates of poverty in dryland zones in six countries, D'Errico and Zezza (2015) estimated a probit model in which a binary poverty variable was regressed on a set of control variables that included household demographic characteristics (a vector $H$ including household size, dependency ratio, and gender of the household head); household assets (a vector $A$ including average education of adult members, land owned, livestock owned, and an index of access to infrastructure); and a set of variables $S$ indicating the distance of the household from school and health facilities. They controlled for the number of income sources to which the household has access in order to capture the extent to which income diversification may be associated with lower probability of being poor. Finally, they controlled for a range of agro-climatic and soil variables $T$. Included in the regressions were the Aridity Index and soil quality (as measured by organic carbon content), to assess whether those are systematically correlated with poverty status. Much of the concern with livelihoods in drylands is associated with the idea that households in drylands are exposed to a higher level of climate hazards compared to the average household. To capture the effect that these hazards may have on welfare, D'Errico and Zezza introduced as additional right-hand side variables the long-term coefficients of variation of maximum temperature and precipitation during the growing season.

The model can be written as: $Pr\,(Y_i = 1 | X_i) = \Phi(X_i\,\beta)$

where $X_i = f\,(H_i, A_i, S_i, T_i, D_i)$, and $\Phi$ is the standard cumulative distribution function. The dependent variable is an indicator set equal to 1 if a household falls below the poverty line, and 0 otherwise. $D$ is a vector of country fixed effects. The subscript $i$ denotes the households. The approach used is fairly standard in country-level poverty analyses, and as always the results should be interpreted as showing correlation but not necessarily causality.

The model results show that better access to land is associated with lower poverty in all aridity zones (except in arid zones, where the coefficient is not statistically significant). Similarly, greater income diversification is correlated with lower poverty across all zones. Somewhat surprisingly, livestock ownership is strongly correlated (negatively) with poverty in non-dryland zones, but the correlation is insignificant in dryland zones. The effect of the infrastructure index appears to decline with the decline in aridity, being 7 times larger in arid areas and 3 times larger in semi-arid areas than in non-dryland areas. This finding suggests that infrastructure investments in dryland areas could have a particularly pronounced effect on reducing poverty.

*(continued next page)*

**Box 2.2** *(continued)*

An interesting aspect of the modeling results relates to the rainfall, temperature, and soil quality attributes. Rainfall variability is associated with significantly higher probability of households being poor in both arid and semi-arid areas, but the coefficient on rainfall variability is not significantly different from zero in non-dryland areas. In contrast, the coefficient on the variability in maximum temperature is not significant for drylands as a whole. Finally, in dryland zones the organic carbon content of the soil, a proxy for soil fertility, appears to be associated with a lower probability of being poor, while no association between poverty and soil fertility is detected for non-dryland areas.

The overall picture that emerges from this analysis is that the quantity and quality of land resources, access to infrastructure, and exposure to variability in rainfall are strongly correlated with poverty. By and large, the correlates of poverty in dryland zones do not appear to be structurally different from the correlates in non-dryland zones.

droughts, other things equal, than a household that derives a small share of its income from drought-affected activities. Sensitivity is determined in large part by past decisions made by the household regarding the nature and mix of its assets and by its livelihood strategy. Changing the nature and mix of assets, as well as the livelihood strategy, is one of the main avenues the household can follow to enhance its resilience.

- **Coping capacity** refers to the ability of a household to mitigate the impact of droughts and other shocks after they occur. Access to financial resources (from its own savings, from friends or relatives, or from social safety nets) can help the household make up for an income shortfall resulting from, for example, a drop in production following a drought. Liquidating productive assets to mitigate the negative impacts of current droughts may reduce the ability of the household to mitigate the impacts of future droughts, that is, it will reduce the household's resilience.

Since it is unlikely that all risks can be avoided by diversifying household assets and altering income-generating activities to reduce exposure to future shocks, resilience-enhancing strategies usually consist of a combination of actions to reduce sensitivity and actions to increase coping capacity.

The methods used to estimate the number of people exposed to, sensitive to, and unable to cope with droughts and other shocks are described in Chapter 4.

The vulnerability (and by extension the resilience) of a given household depends on the combined effect of these three factors. A household is vulnerable when, by virtue of its physical location, livelihood activities, and assets, it is exposed to droughts and other shocks, sensitive to droughts and other shocks, and lacks the capacity to cope effectively when a drought or some other shock occurs. Conversely, a household is resilient when it is not exposed to droughts and other shocks, or is insensitive to droughts and other shocks, or is able to cope effectively when a drought or some other shock occurs. In the aggregate the resilience of a country in the face of droughts and other shocks increases the lower the share of the population exposed, the lower the share of people sensitive, and the greater the share of exposed and sensitive people who are able to cope. Over time, resilience is determined by the interplay of all three of these dimensions.

The approach used in this book takes into account all three dimensions of resilience, considering the current situation in drylands and also projecting how the three dimensions are likely to evolve in future under a number of plausible scenarios. The approach has the advantage of avoiding the pitfalls of defining policies for drylands based on the individual determinants of resilience. For example, when relatively few people have incomes that are so low as to place them below the poverty line, it would be easy to conclude that the coping capacity of the population is relatively high, since most households dispose of enough assets to be able to recover from a drought, should one occur. Based on this reasoning, policy makers might focus on the poverty headcount as a good indicator of vulnerability.

But focusing in this way on a single dimension of resilience could obscure the fact that even though most households dispose of enough assets to recover from a drought, the livelihood strategy that allowed them to accumulate those assets may be very sensitive to droughts. If this is the case, even if households do not suffer from chronic structural poverty, they may still be subject to stochastic transient poverty, as recurrent droughts will cause them to cycle in and out of poverty over time (Barrett and Carter 2013; Carter and Barrett 2006). That being the case, the population at risk should be understood to include not only the people who are poor today, but also the people who risk becoming poor tomorrow because their income is exposed and sensitive to drought and other shocks.

As the Ethiopia experience shows, policies that succeed in lifting some people out of poverty at a particular point in time do not necessarily guarantee that, as a result of subsequent shocks, many of these people will not fall back into poverty. As a result, it makes sense to explore policies and interventions that can address simultaneously all three dimensions of resilience.

## The policy significance of the determinants of resilience

In considering interventions to reduce vulnerability and improve resilience in the drylands, three types of interventions can be distinguished:

1. **Interventions that reduce exposure:** These are interventions that cause households to take actions before a shock occurs so as to avoid the shock,

including by moving to a region in which droughts occur less frequently or less severely. For example, governments can encourage increased mobility among pastoralists, allowing them to move within or between countries to avoid drought hotspots, or they can facilitate migration away from drought-prone zones by supporting the development of growth poles outside of drylands.

2. **Interventions that reduce sensitivity:** These are interventions that cause households to take actions before a shock occurs so as to reduce the effects of the shock when it hits, for example, by diversifying their income sources or adopting more robust production technologies. For example, governments can support the adoption of drought-resistant crop varieties or promote the uptake of irrigation.

3. **Interventions that improve coping capacity:** These are interventions that allow households to take actions after a shock has hit to speed their recovery from the effects of the shock, for example, by selling off animals, drawing down savings from a bank account, or relying on remittances from relatives. Alternatively, governments can provide improved access to social safety nets or enact policies to support the establishment or expansion of insurance markets.

What should the mix be of the three types of interventions, taking into account that the relative merits of each will differ depending on country circumstances? In some countries, it might make sense to focus efforts on increasing coping capacity of vulnerable households, for example, by strengthening social safety nets or introducing affordable private insurance instruments. In other countries, there may be significant scope for reducing the sensitivity of vulnerable households, for example, by supporting the uptake of better livestock and farming technologies. In still other countries, where the fiscal cost of scaling up safety nets is high and the opportunities to make livelihoods less sensitive to shocks are limited, the priority might be to promote alternative livelihood strategies or encourage the movement of vulnerable people away from drylands.

## Shocks affecting drylands

In the context of drylands, four types of shocks warrant attention from policy makers:

1. **Meteorological shocks** can be caused by weather in the short run or by climate change in the long run.

2. **Health shocks** can affect plants, animals, or people.

3. **Price shocks** occur when households are subject to fluctuations in the prices of goods and services that they purchase or sell.

4. **Conflict** can lead to disruption of livelihood activities, loss of property displacement, and/or bodily injury including death.

This book focuses primarily on meteorological shocks, specifically *droughts*, with which vulnerability in the drylands is most often associated. Less attention is paid in the book to the other three types of shocks, each of which has unique causes that call for specialized solutions.

Resilience in the drylands is also affected by longer-term processes that over time undermine livelihood activities, such as *land degradation* and *climate change*. Because the effects of these longer-term processes are gradual, they rarely precipitate immediate humanitarian crises and therefore tend not to attract as much attention. While the impacts of these processes may not be felt immediately, they have the capacity to cause losses at extremely large scale, which is why they are briefly discussed in the following section of the chapter.

## Relationship between resilience and poverty

What is the relationship between resilience and poverty? Poverty reduction remains a high-order objective of development policy; building resilience to shocks is not necessarily a goal in itself, but it is an essential pre-condition for achieving poverty reduction. The reason is that when households and communities are repeatedly hit by shocks and lack the means to respond, they have difficulty accumulating the human, physical, and natural capital needed to lift themselves out of poverty. Increasing resilience will not automatically lead to poverty reduction; for poverty to be reduced, a number of additional actions have to be taken, for example, improving health services, strengthening educational systems, and improving access to markets for inputs and outputs. But even if increasing resilience is not a sufficient condition for poverty reduction, it is a necessary one, because households that are unable to cope with the impacts of drought and other shocks normally will not be able to save enough to augment their endowment of productive assets and increase their potential to generate income.

If building resilience can contribute to poverty reduction, the converse is also true. Reducing poverty can be a way to increase resilience, but reducing poverty does not automatically result in enhanced resilience. Resilience is determined by the three factors described above—exposure, sensitivity, and coping capacity. For purposes of this book, to allow estimation of the numbers of people who are resilient—that is, able to recover from the effects of a shock—the poverty line is used to determine coping capacity: households that following a shock see their income fall below the poverty line are deemed unable to cope (that is, these households are considered non-resilient), whereas households that following a shock see their income remain above the poverty line are deemed able to cope (that is, these households are considered resilient). Whether a given household will see its income fall below the poverty line following the occurrence of a shock depends on the household's income level before the onset of the shock, its degree of exposure to the shock, and the sensitivity of its livelihood strategy to the effects of the shock. Relatively poor households that started out just above the poverty line may be considered resilient if they are not highly exposed to the shock or if their income is not sensitive to the effects of the shock; relatively wealthy households that started out well above the poverty line may be considered non-resilient if they are highly exposed to the shock or if their income is extremely sensitive to the effects of the shock. In summary, poverty influences resilience, but it does not in itself determine resilience, and resilience is an essential component of a strategy to eradicate poverty in a lasting manner.

## Note

1.  This definition focuses on people, not on ecosystems (see box 2.1). For simplicity, the book refers mainly to households, since most data are collected at the household level.

## References

Barrett, C.B., and M.R. Carter. 2013. "The Economics of Poverty Traps and Persistent Poverty: Empirical and Policy Implications." *The Journal of Development Studies* 49(7): 976–90.

Carter, M.R., and C.B. Barrett 2006. "The Economics of Poverty Traps and Persistent Poverty: An Asset-Based Approach." *Journal of Development Studies* 42(2): 178–99.

D'Errico, M., and A. Zezza. 2015. "Livelihoods, Vulnerability, and Resilience in Africa's Drylands: A Profile Based on the Living Standards Measurement Study-Integrated Surveys on Agriculture." Unpublished report. World Bank, Washington, DC.

Holling, C.S. 1973. "Resilience and Stability of Ecological Systems." *Annual Review of Ecology and Systematics* 4: 1–23.

Kerven, C., and R. Behnke, eds. 2014. "Human, Social, Political Dimensions of Resilience." Unpublished paper, FAO, Rome.

Scandizzo, P.L., S. Savastano, F. Alfani, and A. Paolantonio. 2014. "Household Resilience and Participation in Markets: Evidence from Ethiopia Panel Data." Unpublished document, World Bank, Washington DC.

Strzepek, K, N. Strzepek, and J, Neumann. 2014. "Analysis of Drought and Rained Crops in Africa." Report to World Bank. Industrial Economics, Inc., Cambridge, MA.

Wiesmann, D., L. Bassett, T. Benson, and J. Hoddinott. 2009. "Validation of the World Food Programme's Food Consumption Score and Alternative Indicators of Household Food Security." International Food Policy Research Institute (IFPRI) Discussion Paper 00870. IFPRI, Washington, DC.

# Chapter 3

# Vulnerability in Drylands Today

*Raffaello Cervigni, Michael Morris, Pasquale Scandizzo,*
*Sara Savastano, Adriana Paolantonio, Federica Alfani, Alberto Zezza,*
*Zhe Guo, Marco D'Errico, Riccardo Biancalani, Sally Bunning,*
*Monica Petri, Mohamed Manssouri, Carol Kerven, Roy Behnke*

## Quantifying the dimensions of vulnerability across livelihood types

How many people living in dryland zones in East and West Africa are vulnerable? Who are these people, and what are their livelihood strategies? What types of resources are needed by these people to become resilient? And how are the numbers of vulnerable people likely to evolve over the long run as the population grows and the economy transforms?

If these questions are to be answered, *vulnerability* and *resilience* must be defined in a way that makes the two concepts easily measurable. In this book the following definitions are used to arrive at quantitative estimates of the numbers of vulnerable and resilient people living in drylands:

- **People exposed to droughts and other shocks** are defined as people living in dryland areas, that is, areas classified according to the Aridity Index as hyper-arid, arid, semi-arid, or dry subhumid. Because most population data for African countries are not geo-referenced, it was necessary to spatialize UN population data using gridding methods routinely used in the literature. A major source was the dataset developed at the Columbia University Center for International Earth Science Information Network (CIESIN) under the Global-Urban Mapping project (GRUMP) (for details see SEDAC 2015).

- **People sensitive to drought** are defined as the share of people dependent on agriculture, evaluated based on recent International Monetary Fund (IMF) estimates of the employment shares of agriculture (Fox et al. 2013), and assuming that people below working age depend on agriculture in the same proportion as people above working age. All those dependent on agriculture are assumed to be equally sensitive to droughts and other shocks. This is admittedly a simplification, since the income share derived from agriculture varies across households. However, data needed to assess consistently across countries the income share derived from agriculture are not readily available. Survey-based evidence (figure 3.1) suggests that in dryland areas, the share of income coming from farming and livestock-keeping is at least 60 percent of the total, so this assumption should not bias the analysis excessively.

- **People unable to cope with the effects of droughts and other shocks** are defined as the proportion of exposed and sensitive people living below the international poverty line of US$1.25 per person per day. Separate estimates of rural and urban poverty rates are rarely available, so the national (overall) poverty rate was used. The resulting estimates of the number of vulnerable people are undoubtedly conservative, because (1) poverty is usually higher in rural areas than in urban areas and (2) poverty is usually higher in dryland areas than in non-dryland areas.

Recognizing that in drought years, people dependent on agriculture experience income losses, in some of the analyses carried out for this book the number of people unable to cope is estimated using other poverty lines. Based on World Food Programme (WFP) survey evidence, it is assumed that households with incomes exceeding the international poverty line of US$1.25 per person per day by 15 percent, 30 percent, and 45 percent would become unable to cope in the

**Figure 3.1** Income sources, drylands vs. non-drylands, selected countries, 2010 (%)

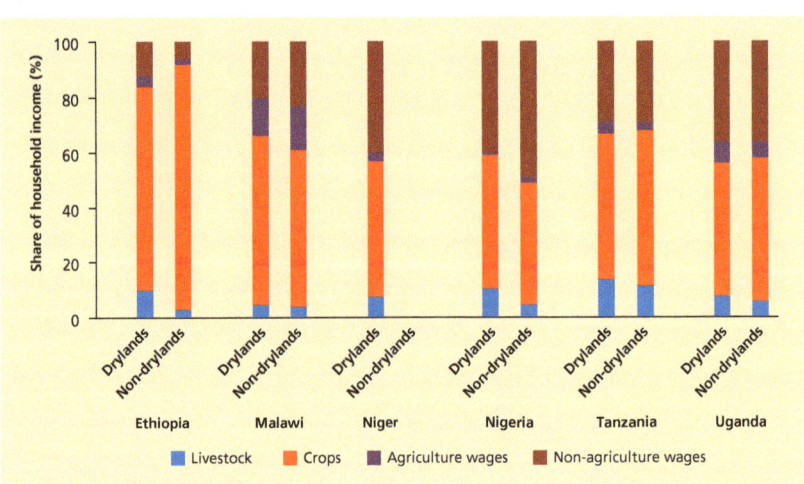

Source: D'Errico and Zezza 2015.

event of mild, moderate, and severe droughts, respectively. In each case, the corresponding poverty headcount is estimated based on income distribution data obtained from the PovCalnet[1] database.

Using the previous definitions it is possible to estimate the dimensions of vulnerability and resilience in the drylands of Africa in the baseline year of 2010 (table 3.1). Throughout the entire region, of the total 424 million people living in drylands (exposed to drought and other shocks), approximately 240 million were dependent on agriculture (sensitive to droughts and other shocks). Of these, some 97 million people were living below the poverty line (unable to cope with droughts and other shocks). In East and West Africa, the two sub-regions that are the main focus of this book, the equivalent numbers were 306 million people exposed, 186 million people sensitive, and 71 million people unable to cope with the effects of droughts and other shocks. Most exposed to droughts and other shocks were the people living in the driest zones, including the hyper-arid, arid, and semi-arid zones. In these three zones, the population unable to cope with the effects of droughts and other shocks was on the order of 46

**Table 3.1** Dimensions of vulnerability in Africa's drylands, 2010 (million people)

| Regions/aridity zones | Exposed | Sensitive | Unable to cope |
|---|---|---|---|
| **East Africa** | **150.6** | **96.6** | **29.2** |
| A. Hyper-arid | 4.7 | 2.9 | 0.5 |
| B. Arid | 30.5 | 18.8 | 3.9 |
| C. Semi-arid | 64.5 | 41.7 | 11.0 |
| D. Dry subhumid | 50.9 | 33.1 | 13.8 |
| **West Africa** | **155.5** | **89.9** | **42.2** |
| A. Hyper-arid | 0.9 | 0.5 | 0.2 |
| B. Arid | 19.2 | 12.2 | 4.8 |
| C. Semi-arid | 90.6 | 53.2 | 26.3 |
| D. Dry subhumid | 44.8 | 23.9 | 11.0 |
| **Subtotal East and West Africa** | **306.1** | **186.4** | **71.5** |
| **Central Africa** | **13.0** | **8.5** | **5.1** |
| B. Arid | 0.1 | 0.1 | 0.0 |
| C. Semi-arid | 3.2 | 1.9 | 0.5 |
| D. Dry subhumid | 9.7 | 6.6 | 4.6 |
| **Southern Africa** | **105.6** | **44.2** | **20.8** |
| A. Hyper-arid | 0.1 | 0.0 | 0.0 |
| B. Arid | 1.8 | 0.5 | 0.2 |
| C. Semi-arid | 56.8 | 20.7 | 7.8 |
| D. Dry subhumid | 47.0 | 23.0 | 12.8 |
| **Grand Total** | **424.7** | **239.2** | **97.3** |

*Source:* Calculation based on the approach discussed in the Appendix.

million people, or roughly 15 percent of the total dryland population in East and West Africa.

Among the people who are exposed, sensitive, and unable to cope, in any given year only some will actually experience a drought or other type of shock. Since the frequency, geographical scale, and severity of shocks is stochastic, this number will vary considerably from year to year. The crop model developed by the African Risk Capacity (ARC) team, in combination with weather data reflecting the historical record of the past 20 years, was used to estimate the average share of people expected to be affected by drought annually. Depending on the country, the average share of people living in drylands expected to be affected by drought in any given year ranges from 7–20 percent, with an overall average of 14 percent (figure 3.2).

The estimated distribution of drought impacts is shown in map 3.1. As discussed later in the book, these figures have particular policy significance because they determine the amount of resources that will have to be committed on a long-term basis to fund social safety nets needed to provide support to all of the people affected by droughts.

**Figure 3.2** Number and percentage of people vulnerable to and affected by drought, selected countries, 2010

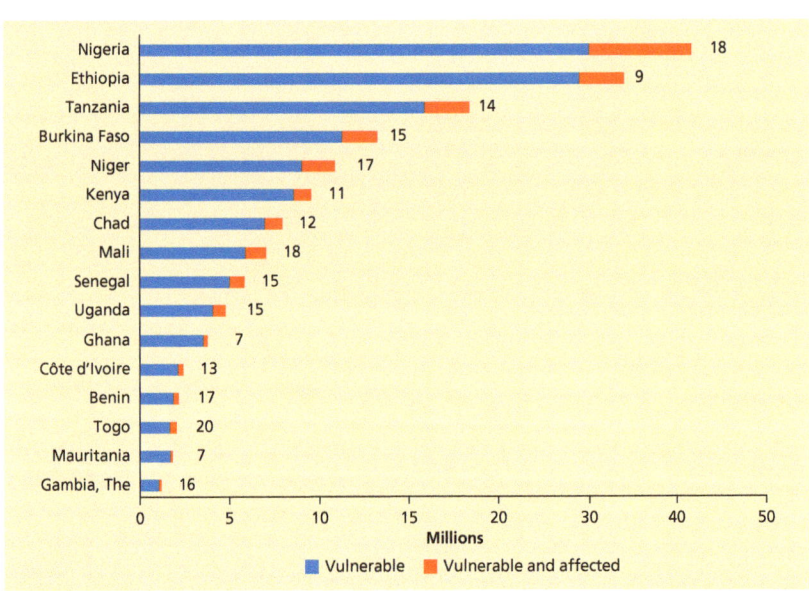

*Source:* Calculation based on the approach discussed in the Appendix.

*Note:* The figures appearing to the right of the bars indicate the average number of drought-affected people, expressed as a percentage of the total number of vulnerable people.

**Map 3.1** Projected number of drought-affected people, annual average, selected countries, 2010

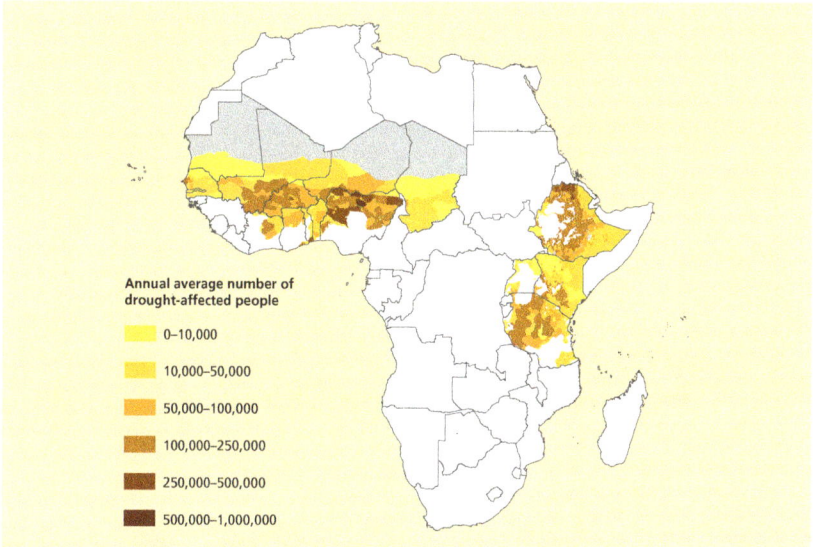

Annual average number of
drought-affected people

0–10,000

10,000–50,000

50,000–100,000

100,000–250,000

250,000–500,000

500,000–1,000,000

*Source:* African Risk Capacity Agency 2015. Reproduced, with permission from Joanna Syroka; further permission required for reuse.

*Note:* Using as a baseline the 2010 population, the map shows the number of vulnerable people in each polygon likely to be affected by drought in a 12-month period. The number of vulnerable people was calculated based on the number of people dependent on agriculture and living below the international poverty line. The number of people likely to be affected by drought was estimated with the help of the ARC model using crop yield simulations (for details, see Carfagna, Cervigni, and Fallavier 2016). Rainfall data for the past 21 seasons, considered to be a representative distribution of the rainfall that could have been experienced in 2010, were used to generate for each polygon 21 estimates of the drought-affected population; these were then used to calculate the annual average (or expected) drought-affected population. The map shows drought "hotspots," identified in terms of the average absolute number of people affected. The average absolute number provides a composite picture of the expected frequency and magnitude of drought events in a given polygon and the number of people considered at risk from drought in that polygon. An increase in either factor will increase the annual average number of drought-affected people in a given polygon.

## Estimating vulnerability across livelihood strategies

Three representative livelihood strategies were identified for use in (1) projecting the likely consequences of the ongoing demographic and socioeconomic transformation of the drylands, and (2) assessing the scope for increasing resilience through the technical interventions. These strategies are:

1.  Livestock-keeping only ("pastoralist households")

2.  Mixed livestock-crop production ("agro-pastoralist households")

3.  Crop production only ("farming households")

In the absence of detailed census data, it is difficult to know exactly how many people are engaged in each of these three livelihood strategies. The approach used for this book was to combine information obtained from socio-economic surveys, mainly those found in the World Bank Survey-based Harmonized Indicators Program (SHIP) database, with estimates from agro-ecological analysis. In particular the calculations were:

- The number of people engaged in crop production only ("farming households") was estimated based on the number of rural households that reported not owning any livestock (in a few countries where data on livestock ownership were not available, expert judgment was used).

- The number of people engaged in livestock production was estimated as the residual (that is, those not engaged in crop production only). To distinguish between people engaged in livestock-keeping only ("pastoralists") and people engaged in mixed livestock-crop production ("agro-pastoralists"), the ILRI/FAO map of livestock production systems was superimposed on the population map. People living in locations associated with livestock-only production systems were assumed to be pastoralists, and people living in locations associated with mixed crop-livestock systems were assumed to be agro-pastoralists. (Details of the calculations appear in De Haan 2016.)

The results of these estimations for East and West Africa are summarized in table 3.2. In 2010, of the approximately 171 million people living in drylands and dependent on agriculture, about 26 million were pastoralists, 105 million were agro-pastoralists, and 40 million were crop farmers.

**Table 3.2** Estimated agriculture-dependent population, East and West Sub-Saharan Africa, 2010 (millions of people)

|  | Population | Dependent on agriculture | of which | | |
|---|---|---|---|---|---|
|  |  |  | Crop farming | Pastoralism | Mixed farming |
| **Drylands** | **247.7** | **171.2** | **39.5** | **26.2** | **105.5** |
| East Africa | 92.2 | 64.7 | 17.6 | 12.7 | 34.3 |
| West Africa | 155.5 | 106.5 | 21.9 | 13.5 | 71.1 |
| **Non-drylands** | **269.0** | **195.7** | **57.3** | **13.0** | **125.4** |
| East Africa | 109.6 | 78.2 | 20.8 | 4.4 | 53.1 |
| West Africa | 159.4 | 117.5 | 36.5 | 8.6 | 72.3 |
| **Total** | **516.7** | **366.9** | **96.8** | **39.3** | **230.8** |

*Source:* Population data from UNFPA (United Nations Population Fund); breakdown by aridity zone from IFPRI (International Food Policy Research Institute).

**Figure 3.3** Estimated dryland population dependent on agriculture in 2010 by country and livelihood type (millions of people)

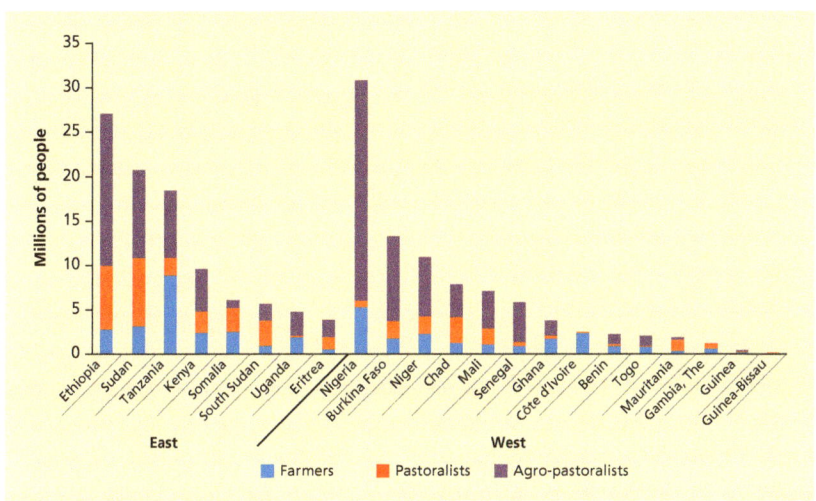

*Source:* Calculation based on the approach discussed in the Appendix.

At the level of individual countries, agro-pastoralists are usually the dominant group, but not always, as the relative importance of the three livelihood strategies varies as a function of local agro-ecological and socioeconomic characteristics (figure 3.3).

Having established basic order-of-magnitude estimates of the determinants of vulnerability among dryland populations, as well as of the distribution of people across main livelihood types, the rest of this chapter discusses key aspects of the development challenge faced by people living in drylands. These relate to natural capital (section on land degradation), physical capital (section on access to infrastructure), and social capital (section on political economy factors affecting resilience).

## Selected drivers of vulnerability

### Land degradation

What is the relationship, if any, between land quality and resilience? More specifically in the context of this book, to what extent does land degradation influence patterns of vulnerability and resilience in dryland regions of Africa?

These seemingly straightforward questions turn out to be difficult to answer for two reasons. First, land quality characteristics are often evaluated differently by different groups of users, making empirical measurement of land quality conceptually challenging. Second, even when there is agreement about how land quality characteristics should be measured, the needed data may be lacking.

Between 2006 and 2011 the Land Degradation Assessment in Drylands Project (LADA)—funded by the Global Environment Facility (GEF), implemented by the United Nations Environment Programme (UNEP), and executed by FAO—created a database and a set of associated analytical tools for use in formulating informed policy advice on land degradation in drylands at global, national, and local levels. Using available global datasets, a Global Land Degradation Information System (GLADIS) was developed that can be used to assess land quality status and trends based on four biophysical parameters (biomass, biodiversity, soil, and water). To avoid the perspective bias described above, an effort was made to maintain a neutral point of view; thus, land quality was evaluated based on all potential uses, rather than in terms of its usefulness for one purpose or another.

As part of this study, information available through GLADIS was used to assess two key characteristics of land in the dryland regions of Africa: land degradation status and land degradation trends. The results of this assessment (reflected in figure 3.4 and map 3.2) generated several important insights:

**Figure 3.4** Shares of total land area in dryland regions by land degradation classes (%)

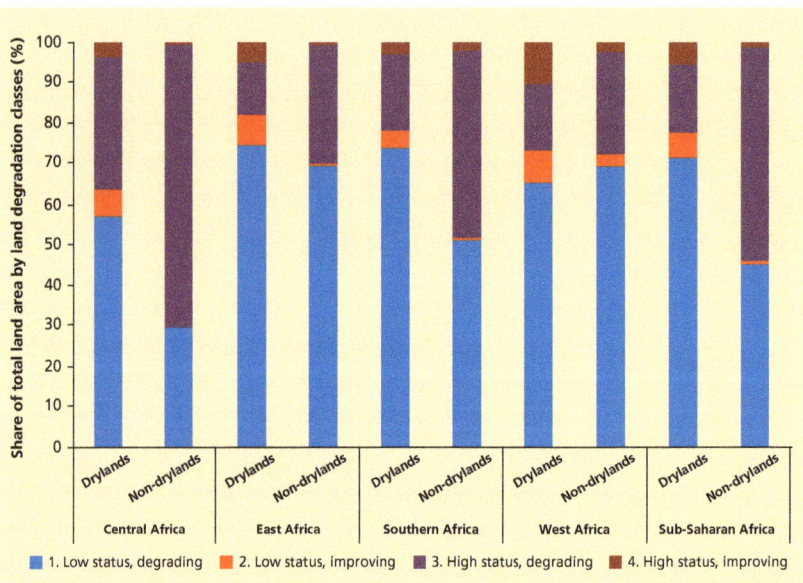

Source: Biancalani, Petri, and Bunning 2015.

**Map 3.2** Land degradation classes, Sub-Saharan Africa (number of people)

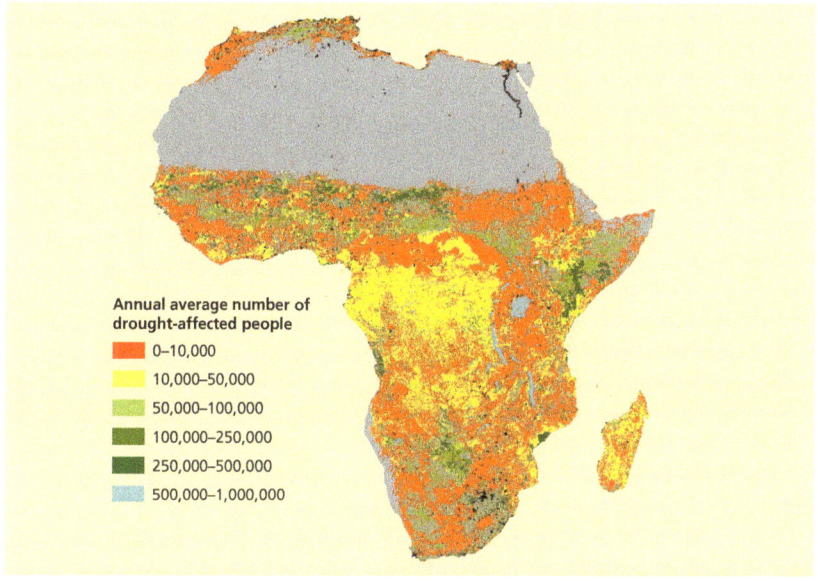

Annual average number of
drought-affected people

- 0–10,000
- 10,000–50,000
- 50,000–100,000
- 100,000–250,000
- 250,000–500,000
- 500,000–1,000,000

*Source:* Biancalani, Petri, and Bunning 2015. Reproduced with permission; further permission required for reuse.

- Much of the land in dryland zones of Africa is currently degraded; on average, the land in dryland zones is more degraded than the land in non-dryland zones.

- Much of the land in dryland zones of Africa is becoming more degraded, but not everywhere. In some locations, land quality is improving, thanks to large-scale land reclamation projects and re-greening efforts.

- In dryland zones of Africa, *land quality status* does not appear to be highly correlated with population density, that is, land is not necessarily more degraded in areas in which population density is highest.

- In dryland zones of Africa, *land quality trends* are highly correlated with population density, that is, land quality is getting worse in areas in which population density is highest.

The productivity and sustainability of the livelihood strategies that currently dominate in the drylands (livestock keeping and crop production) are sensitive to many of the factors included in the land quality indices reported by GLADIS, so the extent of highly degraded land in drylands and the negative trends observed in many locations provide grounds for concern. At the same time, the fact that the trend is positive in quite a few locations in the drylands shows that with the appropriate mix of policies, institutions, and supporting investments, land degradation processes can be slowed and even reversed.

Slowing and reversing land degradation in the drylands is an important priority, with the potential to affect positively the livelihoods of millions of poor and vulnerable households. An even greater priority is promoting the adoption of sustainable land management practices in areas that are still relatively unaffected by degradation and in which the potential of the land is not yet being fully exploited (as evidenced by the existence of large yield gaps in livestock and/or crop production systems). In the latter areas, use of sustainable land management measures could raise productivity while preventing land degradation and increasing resilience of the interested populations.

It is important to stress that resilience is not only affected by land degradation as such. The progressive reduction of land productivity due to degradation processes implies a reduction in income, which in turn increases vulnerability. Implementation of sustainable land management measures, while not without costs, is essential for reversing the vicious circle triggered by land degradation, and for increasing and stabilizing land productivity and contributing to livelihoods and development.

**Map 3.3** Travel time to the nearest town of 100,000 people, dryland zones, 2010 (hours)

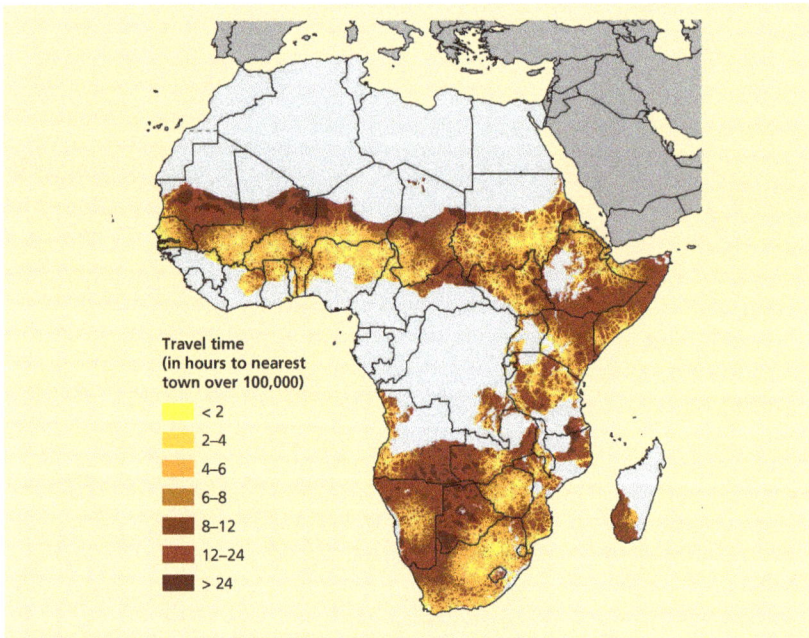

*Source:* World Bank.

## Access to infrastructure

What is the relationship, if any, between isolation and resilience? More specifically, to what extent does a household's ability to access urban centers—home to services and markets—affect vulnerability and resilience in dryland regions of Africa?

The question is important, because many dryland zones are poorly served by transportation infrastructure, and travel times to the nearest large town are extremely high in many areas (map 3.3).

As can be inferred from map 3.3, travel time to the nearest large town increases with the level of aridity. This means that people living in the most arid zones are also the most likely to be disconnected from urban centers (figure 3.5).

The relatively greater degree of isolation of people living in drylands contributes to their vulnerability and lack of resilience. A large body of literature supports the notion that geography matters enormously for economic activities and welfare, with the impacts transmitted mainly through differences in access to markets, access to natural resources, incidence of infectious diseases, and effectiveness of governance (for examples, see Bloom and Sachs 1998; Hentschel et al. 2000; Jalan and Ravallion 2002; Ravallion and Datt 2002). More recently, Stifel and Minten (2008) examined the effects of isolation on agricultural productivity in Madagascar. They discovered a strong inverse relationship between isolation and productivity, which they attributed to (1) transportation-induced

**Figure 3.5** West Africa: Share of population at four hours or more travel from nearest market (%)

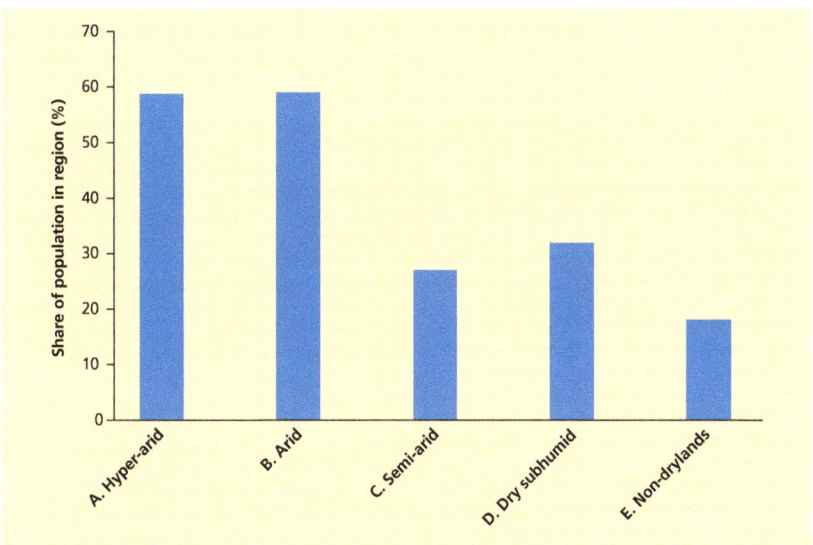

*Source:* World Bank.

transaction costs, (2) the inverse relationship between plot size and productivity, (3) increasing price variability and extensification onto less fertile land, and (4) insecurity. While comparable studies have not yet been done for many dryland regions in Africa, the same factors presumably are at play, as discussed at greater length in Chapter 8.

## Political economy factors affecting resilience

The conceptual framework used in this book to gain insights into the determinants of vulnerability and resilience in drylands considers how existing livelihood strategies may be affected by exogenous shocks, especially droughts. The impacts of these shocks on individual groups in the population may be considerably influenced, positively or negatively, by state policies and programs. In dryland regions of Africa, as nearly everywhere else in the world, state policies and programs are rarely neutral in terms of the costs they impose and the benefits they confer. Designed and implemented by human agents, they tend to favor the interests of groups with sufficient economic and political power to influence the political process. In cases where the interests of all groups in society are well represented, policies and programs can lead to efficient and equitable use of resources, thereby advancing the interests of all. But in cases where state-sanctioned actors are able to exert unchecked power, this may lead to the expropriation of resources, which can exacerbate the vulnerability of dryland populations and undermine their resilience.

This is not just a theoretical matter. In many dryland countries in East and West Africa, uneven distributions of wealth and power combined with differing abilities to influence public policy have resulted in the de facto marginalization of certain groups. Most notable among these are many nomadic pastoral groups, whose ability to engage effectively in political processes often is impeded by their low numbers, peripatetic lifestyle, limited economic power, and lack of integration into mainstream society. The marginalization of many pastoral groups is perpetuated by an internally reinforcing cycle: lacking wealth and power, these groups are not able to make their voices heard in the political dialogue, hence they are not able to gain access to essential resources and services that might allow them to increase their wealth and gain political power, leaving them trapped in poverty and perpetually unable to influence the political process.

The marginalization of many dryland groups can be seen in the skewed distribution of social services, particularly for human health and education. These are often poorly provided in dryland areas, for a number of reasons including insufficiency of national government budgets, distance from national capitals, and high unit costs of provision in areas of low population density (UNDP/ UNCCD 2011). These factors come into play to an even greater extent in pastoral areas, where they are combined with the difficulties of serving mobile

**Figure 3.6** Childhood vaccination coverage in Kenya and Ethiopia, 2005 (%)

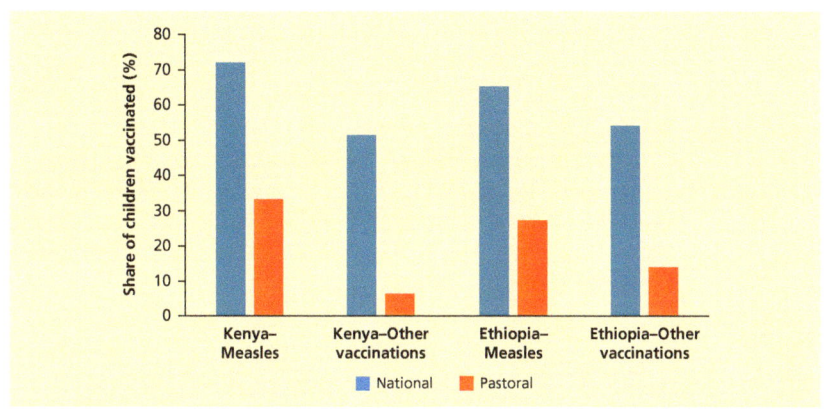

*Source:* Adapted from Ali and Hobson (2005).

populations, further cultural and political marginalization of pastoralists from national mainstreams, and pastoralists' own mistrust of external service providers. The consequences can be dramatic. For example, with respect to health services, dryland areas of Kenya and Ethiopia lag far behind other areas in vaccination coverage for measles and other diseases (figure 3.6).

Similarly with respect to education services, gross enrollment ratios for primary school-age children are low across dryland countries of East Africa, with even lower ratios among pastoral children (figure 3.7). Eighty-one percent of Kenyan adults and 87 percent of Ethiopian adults resident in dryland pastoral

**Figure 3.7** Primary education gross enrollment ratios (GER), IGAD countries, 1999–2001 (%)

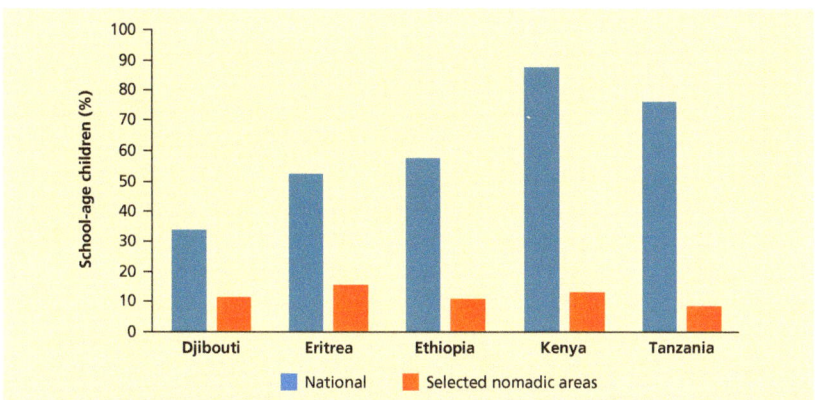

*Source:* Based on data appearing in Morten and Kerven (2013) and Carr-Hill and Peart (2005).
*Note:* IGAD (Intergovernmental Authority on Development) countries are Djibouti, Ethiopia, Somalia, Eritrea, Sudan, South Sudan, Kenya, and Uganda.

areas have received no formal education, which places them in a position of vulnerability when dealing with those more educated and better connected to national political structures. Education facilitates livelihood diversification and resilience to food crises.

These observed disparities in coverage are due to poor public services provision, not to lack of interest or demand for the services by dryland communities. Survey evidence from East Africa indicates that pastoralists rank basic human needs interventions, including health and education provision, as among their most desired development projects (McPeak, Little, and Doss 2012). The strong correlation between formal education, salaried employment, and a secure, diversified livelihood explains their interest in education. Households with a member who has passed through secondary education are more likely to have members in salaried employment, to receive remittances, and to have higher cash income, higher food expenditures, and higher savings. But the benefits of improved education extend beyond the expansion of livelihood opportunities for individuals. Improved education is also required if pastoral communities are to successfully manage their own self-help associations or equip themselves to better defend their ownership of natural resources against commercial or government appropriation. Finally, improved education advances the interests of segments of dryland society—youth and women—that may be disadvantaged in terms of their social or economic standing and, hence, be more vulnerable to risk.

The lower level of social services received by some of the groups living in the drylands, which is reflected in clear discrepancies in many key development indicators, makes it clear that vulnerability and resilience cannot be understood as phenomena with purely technical causes that call for purely technical solutions. If policies and programs are to be effective in attacking the root causes of vulnerability, they need to be designed taking into account the technical, social, and political dimensions of vulnerability and resilience. Although development agencies are often on uneasy ground in dealing with overtly political issues, in order to be effective, interventions will sometimes need to target explicitly marginalized groups who for various reasons may be absent from the policy dialogue.

At the same time, engaging effectively with all groups can be challenging, because the mere act of getting them to participate may not be sufficient. Marginalized peoples, of necessity keen observers of the politics around resource use and control, may not necessarily show their hand in public forums but rather may suspect outsiders of strategic thinking and give strategic answers in response (Browne-Nuñez and Jonker 2008). Development agencies, both national and international, may launch "participatory" consultation and planning processes with the goal of eliciting the needs of marginalized groups, but they may be blinded to the fact that these processes do not always succeed. Meanwhile, well-placed individuals or groups may continue to operate behind the scenes to further improve their position, further distancing the less well-placed from access and control of critical resources.

In some respects, the challenge of bringing marginalized groups into the policy discourse has grown more difficult in recent years as the reach of the global economy has expanded. State agencies and government officials are frequently self-interested players in the commercial developments that are rapidly taking place in the drylands. As recently as a few decades ago, struggles over the control of dryland natural resources revolved for the most part around competing local elements within rural society. This is no longer the case. Globalization, improved transport and communications, the international market value of agricultural commodities, and the increasing presence of the state in rural areas has awakened international interest in dryland resources and has improved the capacity of outside groups to appropriate them. Three essential and valuable natural resources—water, land, and wildlife—have recently become more exposed to external appropriation, leading to increased incidences of dispossession of the rural communities that formerly used them.

While not always recognized by the development community, political considerations such as those described here will surely influence the effectiveness and the distributional impact of the technical interventions that are discussed in the following pages. These considerations will reappear in the concluding chapters when policy implications are discussed, because policies and programs to reduce vulnerability and increase resilience can be designed in ways that strengthen the ability of dryland groups to make their voices heard and hold their governing institutions to account.

## Note

1.  PovCalNet is an online analysis tool for global poverty monitoring maintained by the World Bank Group. See http://iresearch.worldbank.org/PovcalNet/.

## References

Ali, A., and M. Hobson. 2005. "Social Protection in Pastoralist Areas." ODI Humanitarian Protection Group, London.

Biancalani, R., M. Petri, and S. Bunning. 2015. "Land Use, Land Degradation, and Sustainable Land Management in the Drylands of Sub-Saharan Africa." Unpublished paper, FAO, Rome.

Bloom, D., and J. Sachs. 1998. "Geography, Demography and Economic Growth in Africa." *Brookings Papers on Economic Activity* 2: 207–73. Brookings Institution, Washington, DC.

Browne-Nuñez, C., and S.A. Jonker. 2008. "Attitudes Toward Wildlife and Conservation Across Africa: A Review of Survey Research." *Human Dimensions of Wildlife* 13(1): 47–70.

Carfagna, F., R. Cervigni, and P. Fallavier, editors. 2016 (forthcoming). *Mitigating Drought Impacts in Drylands: Quantifying the Potential for Strengthening Crop- and Livestock-Based Livelihoods.* World Bank Studies. Washington, DC: World Bank.

Carr-Hill, R., and E. Peart. 2005. "The Education of Nomadic Peoples in East Africa—Djibouti, Eritrea, Ethiopia, Kenya, Tanzania, and Uganda: Review of Relevant Literature." African Development Bank and UNESCO.

D'Errico, M., and A. Zezza. 2015. "Livelihoods, Vulnerability, and Resilience in Africa's Drylands: A Profile Based on the Living Standards Measurement Study-Integrated Surveys on Agriculture." Unpublished report. World Bank, Washington, DC.

De Haan, C., ed. 2016. *Improved Crop Productivity for Africa's Drylands.* World Bank Studies. Washington, DC: World Bank.

Fox, L., C. Haines, J. Huerta Muñoz, and A. Thomas. 2013. "Africa's Got Work to Do: Employment Prospects in the New Century." International Monetary Fund (IMF) Working Paper 13/201. IMF, Washington, DC.

Hentschel, J., J.O. Lanjouw, P. Lanjouw, and J. Poggi. 2000. "Combining Census and Survey Data to Study Spatial Dimensions of Poverty: A Case Study of Ecuador." *World Bank Economic Review* 14: 147–166.

Jalan, J., and M. Ravallion. 2002. "Geographic Poverty Traps? A Micro Model of Consumption Growth in Rural China." *Journal of Applied Econometrics* 17: 329–46.

Kerven, C., and R. Behnke, eds. 2014. "Human, Social, Political Dimensions of Resilience." Unpublished paper, FAO, Rome.

McPeak, J., P.D. Little, and C. Doss. 2012. *Risk and Change in an African Rural Economy: Livelihoods in Pastoralist Communities.* Routledge ISS Studies in Rural Livelihoods (Book 7). The Hague: Routledge.

Morton, J., and C. Kerven. 2013. "Livelihoods and Basic Service Support in the Drylands of the Horn of Africa." Brief prepared by a Technical Consortium hosted by CGIAR in partnership with the FAO Investment Centre. Technical Consortium Brief 3. International Livestock Research Institute, Nairobi.

Ravallion, M., and G. Datt. 2002. "Why Has Economic Growth Been More Pro-Poor in Some States of India Than Others?" *Journal of Development Economics* 68: 381–400.

SEDAC (Socioeconomics and Data Applications Center). 2015. The Global Urban-Rural Mapping Project (GRUMP). SEDAC, National Aeronautics and Space Administration (NASA), Washington, DC. http://sedac.ciesin.columbia.edu/data/collection/grump-v1.

Stifel, D., and B. Minten. 2008. "Isolation and Agricultural Productivity." *Agricultural Economics* 39: 1–15.

UNDP/ UNCCD (United Nations Development Programme/United Nations Convention to Combat Desertification). 2011. *The Forgotten Billion: MDG Achievement in the Drylands.* New York: United Nations Development Programme; Bonn, Germany: United Nations Convention to Combat Desertification.

## Chapter 4

# Vulnerability in Drylands Tomorrow: Business as Usual Raising Ominous Prospects

*Raffaello Cervigni, Michael Morris, Pierre Fallavier, Zhe Guo,*
*Brent Boehlert, Ken Strzepek*

## Estimating vulnerability in 2030: A scenario modeling approach

An original modeling framework developed expressly for this book (referred to as the *umbrella model*) provides a common analytical framework for integrating findings emerging from the background analysis carried out in different sectors. The umbrella model can be used to project changes in the numbers of vulnerable people living in drylands under a range of scenarios, to evaluate the ability of different interventions to reduce the impacts of droughts, and to estimate the corresponding cost. The umbrella model provides a coherent, albeit simplified, analytical framework that can be used to anticipate the scale of the challenges likely to arise in drylands, as well as generate insights into the opportunities for addressing those challenges.

This chapter briefly summarizes the key elements of the umbrella model (a more detailed description appears in Carfagna, Cervigni, and Fallavier (2016). In addition, it describes the main features of the 2030 business as usual (BAU) baseline scenario, which assumes no interventions are implemented to reduce the number of drought-affected people. Next, Chapters 5 through 11 describe a series of interventions that have the potential to improve the productivity and sustainability of dryland livelihood strategies. Chapter 12 returns to the umbrella model and explores the scope for using these interventions to reduce vulnerability and increase resilience in the drylands.

## A brief description of the umbrella model

To enable comparisons with the 2010 baseline figures presented in Chapter 3, the umbrella model was used to produce 2030 projections of the three components of vulnerability (numbers of people living in drylands who will be exposed to droughts and other shocks, sensitive to droughts and other shocks, and unable to cope with droughts and other shocks):

- **People exposed to droughts and other shocks** are defined as people living in drylands in 2030. The number was obtained by spatializing the UN population projections in accordance with the Global-Urban Mapping Project (GRUMP) dataset used to determine the 2010 baseline. Differences in urban and rural rates of growth are built into the UN projections, reflecting the ongoing trend toward increasing urbanization. Three sets of estimates were generated, one for each of the three UN fertility scenarios (low, medium, high). As with the 2010 baseline, for each scenario the numbers are broken down by aridity class and subnational jurisdiction.

- **People sensitive to droughts and other shocks** are defined as people living in drylands in 2030 and dependent on agriculture. Because economic growth in dryland countries will be accompanied by structural transformation, the share of agricultural employment in total employment is projected to decline; therefore, the umbrella model scales down agricultural employment as a function of economic growth, with the scaling factor derived from a cross-country regression carried out on a large sample of developing countries worldwide. GDP growth per capita in 2030 was calculated for each dryland country by applying to the 2010 baseline growth an increase estimated on the basis of historical GDP growth recorded in each country during the period 1980–2010. To accommodate uncertainty about future GDP growth, three scenarios were modeled (slow, medium, fast), reflecting the 25th, 50th, and 75th percentiles of the distribution of the historical average growth rates (each average in the sample is calculated based on a 20-year period).

- **People unable to cope with the effects of droughts and other shocks** are defined as people living in drylands in 2030 and dependent on agriculture and living below the international poverty line (US$1.25 per day). The number of people living in poverty was calculated by applying to 2030 per capita GDP (estimated as described above) an elasticity coefficient representing the growth elasticity of poverty reduction (GEPR). To accommodate uncertainty regarding the degree to which future growth will result in poverty reduction, three scenarios were modeled: (1) pro-poor growth (GEPR takes on the 75th percentile of the distribution of values observed over the past 20 years); (2) non-pro-poor growth (GEPR takes on the 25th percentile of the distribution observed over the past 20 years; and (3) intermediate case (GEPR fixed at 0.75 for all countries). This approach is designed to capture the overall experience of growth in Africa, which often has not been particularly pro-poor, while avoiding potential distortions that could result if the most recent GDP growth and GEPR values were simply extrapolated (since both parameters may have experienced short-term upward or downward spikes).[1]

## Results: Vulnerability estimates for 2030

Consistent with expectations, under the BAU scenario, exposure, sensitivity, and inability to cope all are projected to grow considerably compared to the 2010 baseline. Important differences can be discerned between countries, however. In addition, the assumptions about future rates of GDP growth and about the impacts of future GDP growth on poverty reduction make a big difference.

The number of people living in drylands who are exposed to droughts and other shocks will grow considerably. Barring an unexpected acceleration in rural-urban migration (that is, beyond the trend already built into the UN population projections), by 2030 the population living in rural areas of the dryland countries is projected to grow between 40 and 120 percent (figure 4.1).

Economic growth will reduce the share of people living in drylands who are sensitive to droughts and other shocks, but probably not fast enough to overcome the effects of demographic growth. As GDP growth generates new employment opportunities in the manufacturing and services sectors, the share of the population living in drylands and dependent on livestock-keeping and crop farming is likely to decrease. Nevertheless, in the presence of rapid population growth, the absolute number of people who depend on these two predominant livelihood strategies and who are exposed and sensitive to droughts and

**Figure 4.1** Projected drylands rural population in 2030 (2010=100, medium fertility scenario)

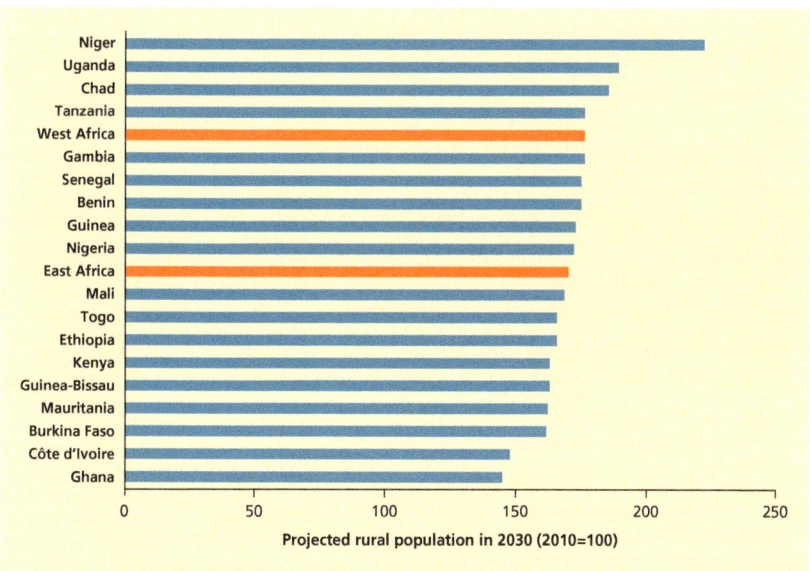

*Source:* United Nations World Population Prospects, 2014 Revision.

**Figure 4.2** People living in drylands projected to be dependent on agriculture in 2030 (2010=100, medium fertility scenario)

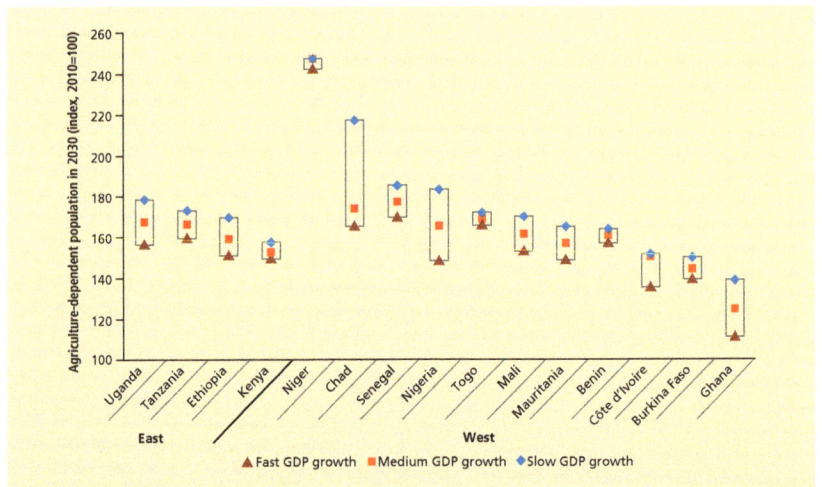

*Source:* Calculation based on the approach discussed in the Appendix.

*Note:* The slow GDP growth scenario is based on the bottom 25 percent of historical growth; the fast GDP growth scenario is based on the top 25 percent of historical growth; and the medium GDP growth scenario assumes a continuation of long-term average historical GDP growth.

other shocks will likely outpace the exits out of agriculture. As a result, the total number of people dependent on agriculture is projected to increase everywhere compared to 2010 levels (figure 4.2).

For many countries, the projected increase falls between 40 and 80 percent, but in a few countries it is much higher (100 percent or more for Chad and Niger). With a few exceptions (Chad and Nigeria), the results are not very sensitive to the assumptions made about future GDP growth.

On aggregate, resilience in drylands will increase only in the presence of growth that is both rapid and more equitable. Three scenarios were considered to explore the likely impacts of different rates of growth and different poverty-reducing effects of growth (figure 4.3). A pessimistic, low-end scenario assumes that growth will be slow and non-pro-poor. An optimistic, high-end scenario assumes that growth will be rapid and pro-poor. An intermediate scenario (used for the rest of the analysis) assumes that growth will be moderate and that the poverty-reducing effect will be modest (GEPR = 0.75). In most countries in East and West Africa, only under the high-end scenario does the number of poor people decrease (signifying an increase in the ability to cope with the effects of drought and other shocks). This result is not universal, however; Niger and Chad are notable exceptions. Under the intermediate scenario, the number of poor people increases significantly (signifying a decrease in the ability to cope with the effects of drought and other shocks). Across the entire set of countries,

**Figure 4.3** Vulnerable people in drylands in 2030 (2010=100, medium fertility scenario)

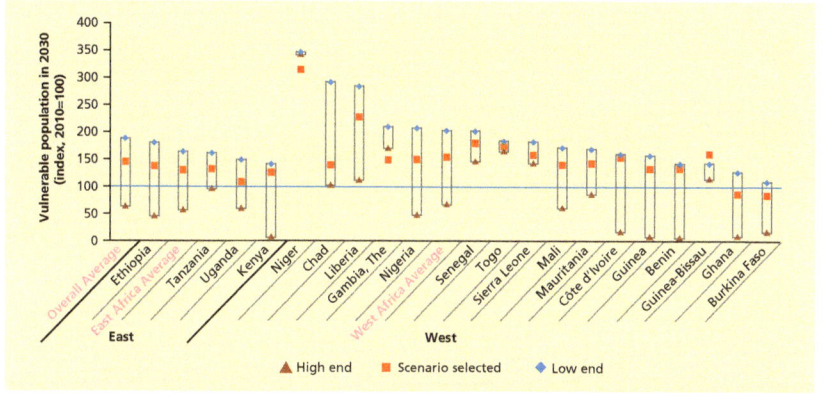

*Source:* Calculation based on the approach discussed in the Appendix.

*Note:* The low-end scenario is characterized by growth that is slow (bottom 25 percent of historical performance) and non-pro-poor (bottom 25 percent of historical performance of the growth elasticity of poverty reduction—GEPR). The high-end scenario is characterized by growth that is fast (top 25 percent of historical performance) and pro-poor (top 25 percent of GEPR distribution). The intermediate scenario selected for the rest of the analysis (reference scenario) is characterized by growth that is modest (equivalent to the long-run historical average) and whose effect on poverty is moderate (GEPR value fixed at 0.75).

the number of poor people increases by 45 percent. The increase is smaller in East Africa (40 percent) compared to West Africa (55 percent). The increase is particularly high in Senegal (80 percent) and Niger (100 percent).

Investment in girls' education can mitigate but not fully address the vulnerability challenge. Investment in the education of girls has been shown to lower fertility rates over the medium to long term (Summers 1992; UNESCO 2011). As fertility rates fall, so does the number of people who are likely to need access to safety nets.

In the drylands, the impact of reducing fertility rates, while non-negligible, is likely to be limited, however. Using the UN low fertility population projections as a first-order approximation of the effects of fertility reduction policies, the increase by 2030 in the number of people vulnerable to shocks could be reduced by 45 percent to 30 percent (figure 4.4).

These sobering results underline the enormity of the challenge facing African governments and the development community more widely. They point to the importance of assessing the ability of different types of interventions to increase the resilience of the poorer segments of the dryland population.

## Effects of climate change on future vulnerability

The BAU projections generated using the umbrella model do not take into account one factor that could significantly affect the calculus of vulnerability and resilience in Sub-Saharan Africa. That factor is climate change. The extent, rate, and likely consequences of climate change are difficult to predict with confidence,

**Figure 4.4** Vulnerable people in drylands in 2030 (2010=100, different fertility scenarios)

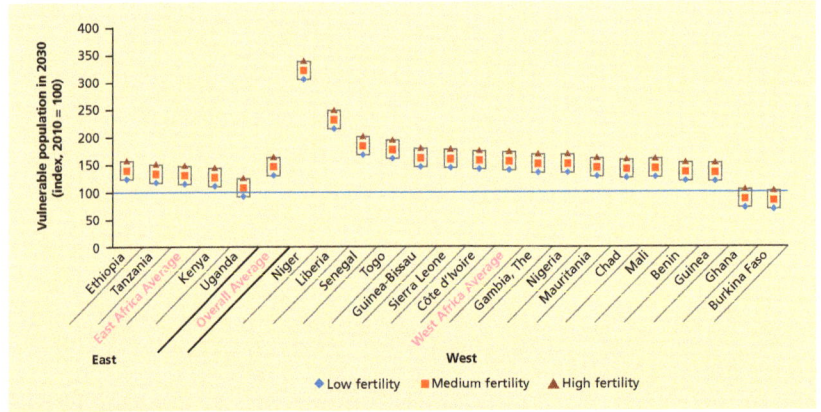

*Source:* Calculation based on the approach discussed in the Appendix.
*Note:* The growth scenario considered is based on an average GDP growth and a fixed 0.75 GEPR value.

and there are considerable differences between the projections made by the leading climate models, but the preponderance of evidence suggests that climate change is likely to have significant impacts worldwide. In Sub-Saharan Africa those impacts are likely to include shifts in the distribution of drylands and expansion in their size, as well as increases in the frequency and severity of extreme weather events experienced within drylands. Under scenarios of faster warming and more pronounced drying, by 2050 the size of drylands in East and West Africa could increase by as much as 40 percent (map 4.1).

These projections suggest that by 2050, climate change could exacerbate the challenges posed by drylands, compounding the effects of rapid population increases and modest growth. It is important to note, however, that climate models do not always agree, particularly in terms of the effects of climate change on precipitation. There is considerable uncertainty not only about the magnitude of the coming changes but also about the direction. To get a fuller picture of the range of possible outcomes, a wide range of scenarios was analyzed to evaluate the impacts on the extent of dryland areas. (box 4.1).

The conclusion is that in some scenarios where wetter conditions are projected to prevail, drylands could actually shrink in size, reducing by as much as 30 percent the population in East and West Africa living in drylands (figure 4.5).

Since the time horizon considered in this analysis is 2030 (when many of the projected effects of climate change may not yet have materialized), for the most part historical weather patterns were used in assessing the effects of droughts on vulnerability and resilience. The fact that longer-term effects of climate change are not explicitly incorporated in the analysis does not diminish the

**Map 4.1** Shift and expansion by 2050 of dryland areas due to climate change (high-end scenario)

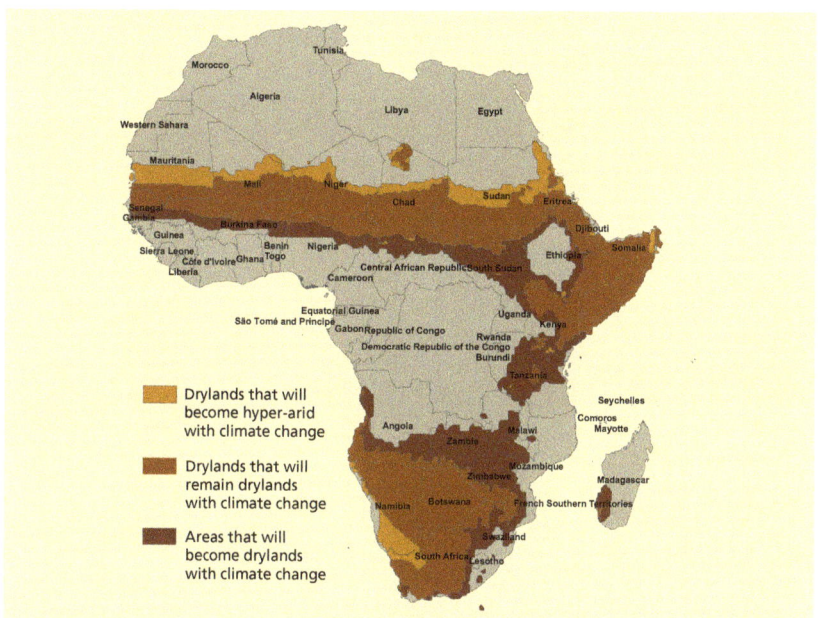

*Source:* Estimates based on general circulation model (GCM) outputs from the CMIP5 ensemble (used in IPCC's 5th Assessment Report).

*Note:* The map shows the extent to which drylands (defined to include all zones with an aridity index between 0.05 and 0.65) could shift and expand by 2050 as a result of climate change. To visualize the largest possible impacts, the map reflects the fastest growth of greenhouse gas (GHG) concentration (RCP 8.5) under the driest of a set of over 99 climate scenarios.

validity of the findings and recommendations, however, because the resilience interventions discussed in subsequent chapters can be instrumental in building resilience not only with the current climate but also with the (probably) much harsher climate of the future. The additional benefits of some of the interventions in the face of climate change are explicitly assessed in Chapter 5, which analyzes the impacts of interventions designed to enhance the productivity and stability of livestock production systems under scenarios featuring more frequent droughts, more severe droughts, or both, than have been experienced to date.

## Note

1.  Estimation errors are particularly likely when poverty rates are interpolated over survey periods: a frequent occurrence for several countries in the PovCalnet database.

## BOX 4.1

# Methodology for projecting shifts in dryland areas under climate change

This chapter's projections of the spread of drylands across Africa under a changing climate rely on a series of calculations that use projections of future climate to predict aridity across Africa at a fine geospatial scale. These projections are based on 99 climate scenarios, each of which is generated from the combination of a general circulation model (GCM) of global climate and a scenario of future greenhouse gas emissions. Thus, 56 of these GCM-emissions combinations use 22 GCMs driven by three Special Report Emissions Scenarios, first adopted in 2000 for the Intergovernmental Panel on Climate Change's (IPCC) Third Assessment, and 43 of these GCM-emissions combinations use 23 GCMs driven by RCP4.5 and RCP8.5, medium- and high-emissions scenarios from the Representative Concentration Pathways (RCPs) adopted for the IPCC's Fifth Assessment in 2013. The primary results of these 99 climate scenarios were then bias-corrected and spatially downscaled, incorporating quantile mapping to account for GCM biases in rainfall intensity distributions. In general, bias-correction spatial disaggregation (BCSD) projections show strong agreement with GCM-projected changes on a large scale and are useful as inputs for impact modeling, particularly in hydrology and agriculture sector work. Each of these BCSD climate projections yielded a time-transient time-series of rainfall and temperature at a 0.5-by-0.5 degree grid across Africa for 2001–50.

Using these climate projections, an aridity index was calculated in a 0.5-by-0.5 degree grid across Africa for 2001–50. This measure of future aridity was then compared to aridity index values calculated for a baseline period from 1961–90 using observed climate data. While measures of drought are designed to identify dry conditions that are temporary aberrations from normal climatic conditions, this measure of aridity identifies regions where low precipitation is the norm. Here, the aridity index is defined simply as annual precipitation divided by annual potential evapotranspiration (PET), where PET is calculated using the modified Hargreaves approach. The Hargreaves approach for calculating PET, which is a function of latitude, average temperature, temperature range, and precipitation, is a preferable alternative to the Penman-Montieth calculation method because it is less data-intensive and proved less likely to underestimate PET in preliminary analysis. Furthermore, the Hargreaves approach has shown greater accuracy than comparable models in previous studies, and the Consultative Group for International Agricultural Research (CGIAR) uses the modified Hargreaves method in its global aridity and PET database.

These baseline and projected aridity indices were then used to predict the shift and expansion of drylands across Africa by 2050, as shown in map 4.1. For this analysis, drylands were defined as areas with an aridity index between 0.05 and 0.65.

**Figure 4.5** Number of people living in drylands in 2050 under different climate change scenarios (2010=100)

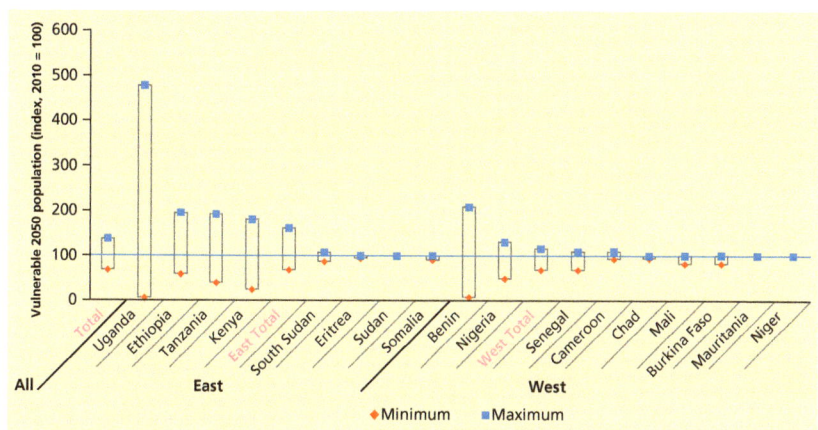

*Source:* Calculation based on the approach discussed in the Appendix.

*Note:* The figure shows how climate change could affect the numbers of people living in drylands in 2050 compared to the 2010 baseline. Values below 100 result from a projected contraction of drylands by 2050; values above 100 result from a projected expansion of drylands by 2050. The figures were estimated using the highest GHG concentration pathway (RCP 8.5). Within each country, the range of values reflects differences between climate models in projected temperatures and precipitation levels, which drive the aridity index.

## References

Carfagna, F., R. Cervigni, and P. Fallavier, eds. 2016 (forthcoming). *Mitigating Drought Impacts in Drylands: Quantifying the Potential for Strengthening Crop- and Livestock-Based Livelihoods.* World Bank Studies. Washington, DC: World Bank.

Summers, L.H. 1992. "Investing in All the People." World Bank Policy Research Working Paper.World Bank, Washington, DC.

UNESCO (United Nations Educational, Scientific, and Cultural Organization). 2011. "Education Counts: Towards the Millennium Development Goals." UNESCO, Paris. http://unesdoc.unesco.org/images/0019/001902/190214e.pdf.

# Part B

# Identifying Solutions

## Chapter 5

# Livestock Production Systems: Seizing the Opportunities for Pastoralists and Agro-Pastoralists

*Cees De Haan, Tim Robinson, Giulia Conchedda, Polly Ericksen, Mohammed Said, Lance Robinson, Fiona Flintan, Alexandra Shaw, Shem Kifugo, Abdrahmane Wane, Ibra Touré, Alexandre Ickowicz, Christian Corniaux, Jill Barr, Cecile Martignac, Andrew Mude, Raffaello Cervigni, Michael Morris, Anne Mottet, Pierre Gerber, Siwa Msangi, Matthieu Lesnoff, Frederic Ham, Erwan Filliol, Kidus Nigussie, Adriana Paolantonio, Federica Alfani*

## Current situation

Livestock-keeping is one of the most important livelihood activities practiced in the drylands of Africa. In the countries of East and West Africa in which drylands are important, the livestock sector is economically significant, with production of meat and milk typically comprising 5–15 percent of total GDP and up to 60 percent of agricultural GDP. The direct contribution of livestock to GDP is amplified when the indirect benefits of livestock-keeping are factored in, such as production of organic fertilizer and provision of animal traction services. In addition, the livestock sector can be an important earner of foreign exchange, as millions of sheep are shipped every year from the Horn of Africa to the Gulf States, and more than one million head of cattle are trekked or trucked from the Sahel to coastal countries in West Africa. Significantly, with per capita incomes continuing to rise in Sub-Saharan Africa and with wealthier consumers turning increasingly to animal-source foods, regional demand for meat and milk is expected to double by 2030.

Livestock-keeping is the principal livelihood source for 40 million people in the Horn of Africa and the Sahel, and it provides a significant share of income for an additional 40 million people in the two regions. The way in which livestock-keeping contributes to the livelihoods of individual households varies depending on the production system. Two main livestock production systems can be distinguished:

1. **Pastoral systems:** Found mainly in more arid zones (Aridity Index 0.05–0.20), pastoral systems are systems in which livestock-keepers derive the majority of their income from animals that graze natural vegetation, the nutritional value and spatio-temporal distribution of which depend on the variability and intensity of annual precipitation. In pastoral zones, where the potential for crop growth is limited by moisture availability, raising livestock is often the only viable form of agriculture. In pastoral systems, cattle, camels, sheep, and goats are moved around to take advantage of patchy seasonal vegetation. The pastoral system represents a complex form of natural resource management and embodies a finely honed symbiotic relationship between local ecology, domesticated livestock, and people in resource-scarce, climatically marginal, and often highly variable conditions. As explained by Pratt, Le Gall, and De Haan (1997), pastoral systems involve interactions between three different systems in which pastoral people operate, namely the natural resource system, the resource users system, and the larger geopolitical system.

2. **Agro-pastoral systems:** Found mainly in semi-arid zones (Aridity Index 0.2–0.5) and subhumid zones (Aridity Index 0.5–0.65), agro-pastoral systems are systems in which livestock-keepers derive one-half or more of their agricultural income from crop farming and in which crop residues make up an important share of livestock rations (usually 10 percent or more). In semi-arid zones, cattle typically perform multiple roles; in addition to producing meat and milk, they contribute to increased crop productivity by providing draft power and manure, while at the same time converting organic material not suitable for human consumption into high-value food and nonfood products. Agro-pastoral systems also represent a complex form of natural resource management that allows efficient exploitation of a limited and highly variable natural resource base.

The distinction between pastoralists and agro-pastoralists, once quite clear, is becoming increasingly blurred, as pastoralists are increasingly engaging in opportunistic planting of small plots in wetter areas or years as a diversification strategy to complement their livestock production activities.

Over the past four decades, livestock numbers have increased rapidly in the drylands (figure 5.1). Between 1980 and 2010 the livestock population in drylands (expressed in Tropical Livestock Units, TLU)[1] grew at an annual rate of about 3.5 percent per year, faster than the human population in these areas, which grew by about 2 percent per year during the same period. Thus on average the herd/flock size per household and per pastoralist have gone up.

Livestock ownership in the drylands is highly skewed. Based on World Bank Harmonized Household Surveys (SHIP) data and rural Gini coefficients, it is estimated that the wealthiest 1 percent of livestock-keepers own between 9 percent and 28 percent of all animals. The regional averages mask important differences between

**Figure 5.1** Growth of livestock numbers and rural human population in drylands, 1960–2010 (millions)

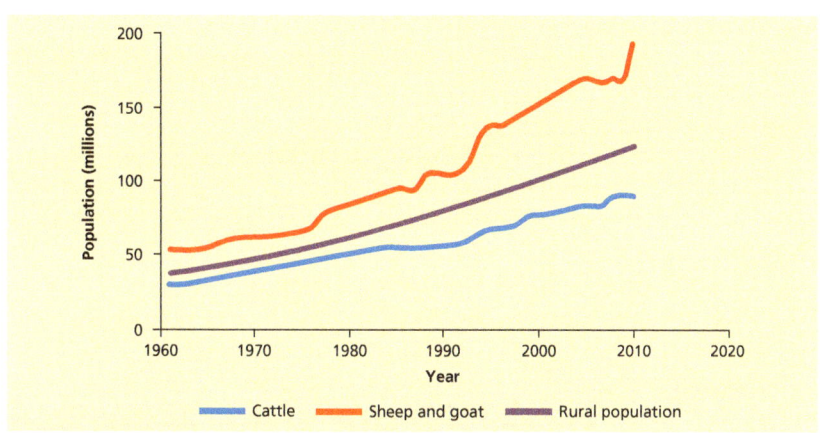

*Source:* FAOSTAT 2015.

regions and among species, however, and they do not reflect changes taking place in the composition of the livestock population. For example, Desta and Coppock (2004)—also mentioned in a report by Headey et al. (2014)—report that in many areas in Ethiopia and Kenya covered by the USAID-funded Pastoral Risk Management (Parima) project, the cattle herd has declined, probably as the result of a series of droughts that reduced herd sizes below the minimum level needed to recuperate.

The vast majority of livestock-keepers in dryland regions of Africa are poor. Estimates reported in the literature, supported by modeling carried out as part of this study, suggest that about 3.5 TLU per capita are needed to meet the basic needs of a typical pastoralist household; the number can be half that much for the typical agro-pastoralist household that is able to supplement income from animals with income from cropping activities. In Sub-Saharan Africa, most households that keep livestock do not have anywhere near that many animals. The estimated 40 million pastoralist livestock-keepers in Africa hold about 51 million TLU (equivalent to 1.3 TLU per capita), and the estimated 80 million agro-pastoral livestock-keepers hold an estimated 76 million TLU (equivalent to less than 1 TLU per capita). Based on these regional aggregates, in the drylands of Africa the "average" pastoral household of six people owns about 6 cattle, 15 sheep, and 15 goats, from which they harvest about 300 liters of milk per year (mostly destined for home consumption), while selling one cow every two years and 10 small ruminants per year. These activities generate about US$700 per year in household income (milk included), or just over US$100 per year per household member. As these numbers show, the "average" livestock-keeper in the drylands of Africa lives below the poverty line.

Livestock-keepers in the drylands of Africa are not only poor, they also face a highly variable environment that exposes them to a variety of shocks from which they may have difficulty recovering.

The most frequent shocks affecting livestock systems in the drylands are undoubtedly extreme weather events, especially periods of severe and prolonged drought. In the Sahel region, the two major droughts that occurred in the 1970s and 1980s led to the deaths of about one-third of all cattle, sheep, and goats (Derrick 1977; Lesnoff, Corniaux, and Hiernaux 2012). Also in the Sahel region the relatively mild drought that lasted from 2010 to 2012 caused about 12 million people to be food insecure (Oxfam 2012). In the Horn of Africa the livestock sector experienced five major droughts between 1998 and 2011, which killed more than one-half of the cattle in the most heavily affected areas and decimated the livelihoods of 3–12 million people (depending on the year).

In addition to being exposed to weather-related shocks, livestock-keepers in many dryland regions of Africa are vulnerable to the effects of conflict. During the past decade alone, episodes of social unrest and civil conflict have broken out in Ethiopia, Kenya, Sudan, South Sudan, Chad, Central African Republic, Niger, Mali, and Nigeria, among other countries, leading to the displacement of millions of people and extensive losses of property, including livestock.

Finally, dryland regions in Africa are particularly susceptible to the increasing criminality that has been linked to the drug and weapons trades, ransom seeking, and the rise of religious extremism. Criminality has destabilized large parts of the Sahel region and the Horn of Africa, displacing many dryland populations, destroying social infrastructure, disrupting traditional livelihood activities, and discouraging tourism (De Haan et al. 2014).

## Opportunities

In considering the prospects for livestock production systems in dryland regions of Africa, it is important not to lose sight of the potential of the sector. Livestock systems in many dryland countries have come under pressure in recent years, resulting in uneven performance, but there is scope for increasing productivity and production. Policy reforms and supporting investments could stimulate changes in production technologies and management practices that could halve the regional deficit in livestock-sourced products that is projected to develop by 2030, should current supply and demand trends continue. At the same time, it is important to recognize that even with these interventions, there will not be enough water, grazing resources, and animals to provide all livestock-keepers in the drylands with an income above the poverty line.

With respect to pastoralism, studies have consistently confirmed the productive efficiency of well-managed pastoral systems in the drylands of Africa, compared, for example, to ranching systems in similarly dry regions in developed countries, including Australia and the United States (see Breman and de Wit 1983). The main opportunities in African pastoral systems, therefore, lie not so much in further increasing productive efficiency, but rather putting in place systems that will enable buffers and rapid adjustments to the "boom and bust" cycles that characterize the system. This could be achieved by maintaining the mobility of herds to allow them to avoid climate shocks, improving animal health services to reduce losses from disease outbreaks and climate shocks; facilitating early destocking when drought is imminent and restocking when rains resume; fostering better market integration, in particular by exploiting complementarities between drylands as the breeding areas and higher rainfall areas for fattening younger stock from the drier areas; and consolidating small holdings of livestock into larger, more resilient, and more viable units.

With respect to agro-pastoralism, the main opportunities lie in the intensification of production systems so as to increase the volume and value of commercial sales. This could be achieved by improving animal genetics to accelerate growth and increase offtake rates, improving animal health services to reduce losses from disease outbreaks and climate shocks, exploiting complementarities between crop and livestock production systems to improve the quantity and quality of available feed resources, and strengthening livestock value chains to increase marketing opportunities. As in the case of pastoralism, consolidation of small herds into larger holdings is needed to ensure that livestock-dependent households have at least the minimum number of animals needed to remain resilient.[2]

To what extent could currently available technologies improve the resilience of livestock-dependent populations living in dryland regions of Africa? To answer this question, it would be important first to understand what would likely happen in the absence of any interventions. The umbrella model (described in Chapter 4) was used to project the numbers of livestock-dependent households likely to be living in the dryland regions of Africa by 2030. Under the business as usual (BAU) scenario, 77 percent of pastoralist households and 58 percent of agro-pastoralist households are projected to own fewer than 5 TLU (figure 5.2). Expressed as a share of livestock-dependent households, the number of poor/vulnerable households is especially high in Niger.

With the BAU baseline established, the potential impacts of four interventions were modeled: (1) improving animal health services, (2) improving access to feed resources, (3) promoting off-take of young male animals from the drylands for fattening in higher rainfall areas, and (4) introducing progressive taxation policies to bring about a more equitable distribution of livestock ownership (box 5.1).

**Figure 5.2** Livestock-keeping dryland households likely to be forced to seek alternative livelihood strategies under a BAU scenario, selected countries, 2030 (%)

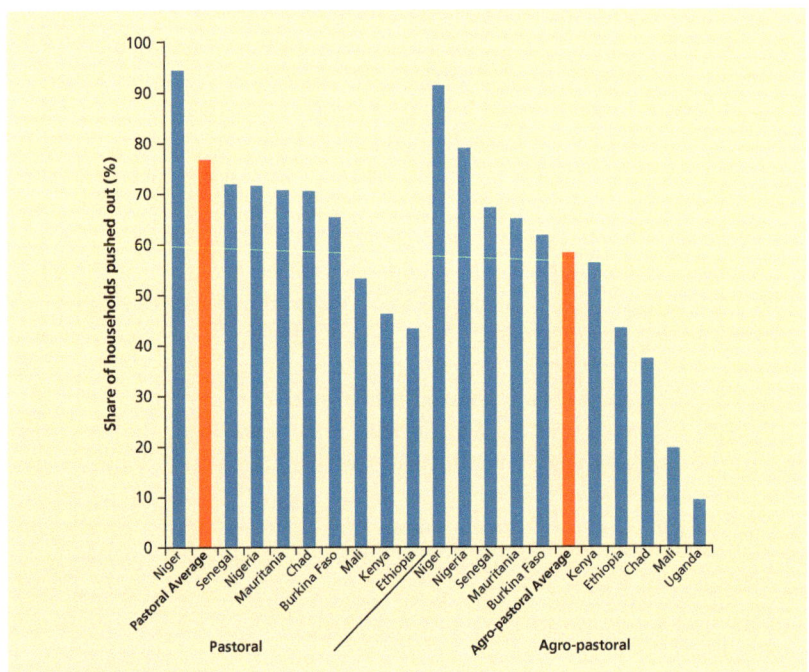

*Source:* De Haan 2016.

## Modeling livestock systems in the drylands

An important original contribution of the study whose results are reported in this book has been to break new methodological ground in the modeling of livestock systems in the drylands. Five simulation models were used in combination to estimate the impacts of the resilience-enhancing interventions on feed balances, livestock production, and household income resilience, under a range of climate scenarios.

1. The **BIOGENERATOR model** developed by Action Contre la Faim (ACF) uses NDVI (Normalized Difference Vegetation Index) and DMP (Dry Matter Productivity) data collected since 1998 by the Satellite pour l'Observation de la Tierre (SPOT) satellite imaging system (Ham and Fillol 2011). The model was used to estimate spatially referenced usable biomass in the drylands (e.g., biomass that is edible by livestock).

*(continued next page)*

**Box 5.1** *(continued)*

2. The **Global Livestock Environmental Assessment Model (GLEAM)** developed by Gerber et al. (2013) calculates at pixel and aggregate level: (1) crop by-products and usable crop residues; (2) livestock rations for different species of animals and production systems, assuming animal requirements are first met by high-value feed components (crop byproducts if given, and crop residues), and then by natural vegetation; (3) feed balances at pixel and aggregate level, assuming no mobility at pixel level and full mobility at grazing shed level; and (4) greenhouse gas (GHG) emission intensity.

3. The **IMPACT model** developed by the International Food Policy Research Institute (IFPRI) is a partial equilibrium global agriculture sector model that can be used to generate baseline projections of agricultural commodity supply, demand, trade, prices, and malnutrition outcomes. On the basis of the feed rations provided by GLEAM, the IMPACT model was used to calculate the production in drylands of meat and milk and to estimate how production will affect overall supply of and demand for these products in the region.

4. The **CIRAD/MMAGE model** consists of a set of functions for simulating dynamics and production of animal or human populations, categorized by sex and age class. It was used to calculate the sex and age distribution of the four main ruminant species (cattle, camels, sheep, and goats), the feed requirements in dry matter, and milk and meat production.

5. The **ECO-RUM model** developed by the Agricultural Research for Development (CIRAD) under the umbrella of the African Livestock Platform (ALive) is an Excel-supported herd dynamics model based on the earlier ILRI/CIRAD DYNMOD. The model was used to estimate the socioeconomic effects of changes in the technical parameters of the flock or herd (e.g., return on investments, income, and contribution to food security).

The modeling exercise benefitted from livestock distribution data contained in the Gridded Livestock of the World (GLW) database (Wint and Robinson 2007) and its most recent update GLW 2.0 (Robinson et al. 2014). It was also informed by information and analysis produced by the FAO livestock supply/demand model (Robinson and Pozzi 2011). For details, see De Haan (2016).

The results of the above models were used as inputs for the final step of the analysis, namely the assessment of the number of households resilient, vulnerable to shocks, and likely to move out of livestock-based livelihoods. These groups were estimated as households owning livestock above or below critical TLU thresholds. The value of these thresholds was estimated using ECO-RUM; and the corresponding population shares were calculated using a log-normal estimate of the TLU distribution, which approximates quite well actual TLU distributions emerging from survey data (SHIP database). The interrelationships between model components as determined by the final analysis are shown in figure B5.1.1.

*(continued next page)*

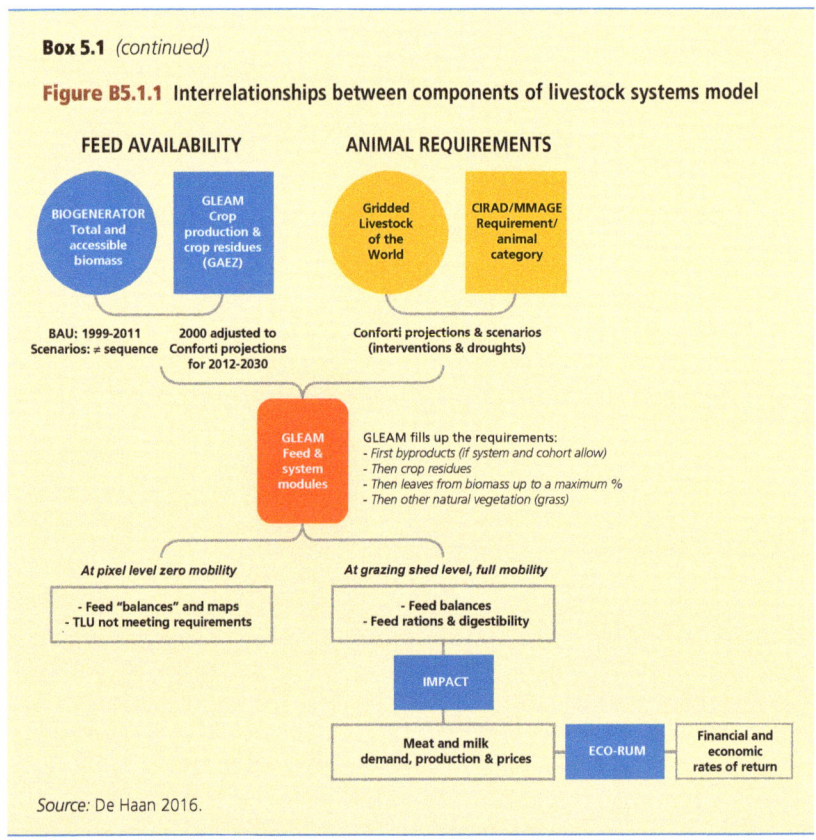

**Box 5.1** *(continued)*

**Figure B5.1.1** Interrelationships between components of livestock systems model

*Source:* De Haan 2016.

The relevance and likely effectiveness of these interventions differ according to the situation, because they address different determinants of vulnerability and resilience.

## Reducing exposure to shocks

Livestock-keepers living in drylands can avoid being affected by shocks, particularly weather shocks, if they can move out of harm's way before the shocks appear. In dryland regions of Africa, and particularly in more arid zones within the drylands, mobile pastoralist livestock systems are generally more productive than sedentary livestock systems precisely for this reason (Catley, Lind, and Scoones 2012; Niamir-Fuller 1999). Drawing on inherited knowledge that has been accumulated over many generations, plus their own personal experience, pastoralists are extremely skilled at moving their animals to take advantage of seasonal feed and water resources while avoiding locations during periods when weather-related shocks are likely to occur. Map 5.1 demonstrates, under a no-drought scenario, the areas in

**Map 5.1** Estimated need for movement of animals in relation to feed, Sahel and Horn of Africa (baseline, no-drought scenario)

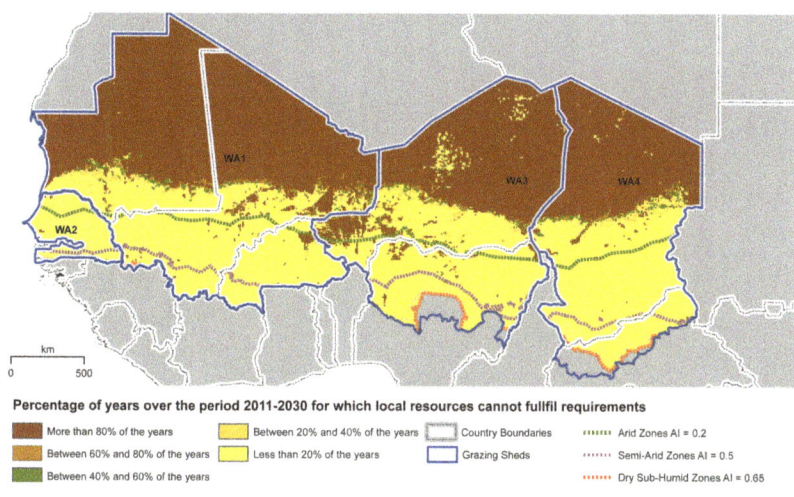

Percentage of years over the period 2011-2030 for which local resources cannot fullfil requirements

| | | | |
|---|---|---|---|
| More than 80% of the years | Between 20% and 40% of the years | Country Boundaries | Arid Zones AI = 0.2 |
| Between 60% and 80% of the years | Less than 20% of the years | Grazing Sheds | Semi-Arid Zones AI = 0.5 |
| Between 40% and 60% of the years | | | Dry Sub-Humid Zones AI = 0.65 |

*Source:* De Haan 2016.

*Note:* WA1, WA2, WA3, and WA4 are labels used to identify the West Africa "grazing sheds." These are defined as areas likely to be used for transhumance predominantly by the same population and herds/flocks each year. The boundaries of the grazing sheds are based on animal mobility patterns known in the literature (SIPSA 2012) and complemented by experts' consultation.

which the local feed resources will be insufficient to provide feed on a year-round basis and for which mobility is essential (these areas appear in orange and red, depending on the frequency with which feed shortfalls occur).

Because mobility is critical, especially for pastoralists, interventions that contribute to improved mobility of livestock-keepers and their animals have the potential to significantly improve the performance of livestock systems in the drylands. Such interventions include: (1) development of water resources to allow better access to underexploited rangelands, (2) organization of feed markets to improve availability of feed in remote areas, and (3) introduction into land use planning of measures designed to facilitate movement of herds and flocks (e.g., through designation of dedicated migration corridors and dry season grazing areas). By improving access to feed, such measures designed to improve mobility can have a large impact on resilience. Figure 5.3 shows how the ratio of resilient households to vulnerable households to nonviable households changes with increasing access to feed.

Other interventions not considered in the modeling exercise can also play an important role in reducing exposure to shocks, including the following: (1) implementation of conflict resolution mechanisms in areas in which livestock-keeping competes with other livelihood activities, to ensure cooperative

**Figure 5.3** Impact of accessibility of feed on the resilience status of livestock-keeping households, share of households (%)

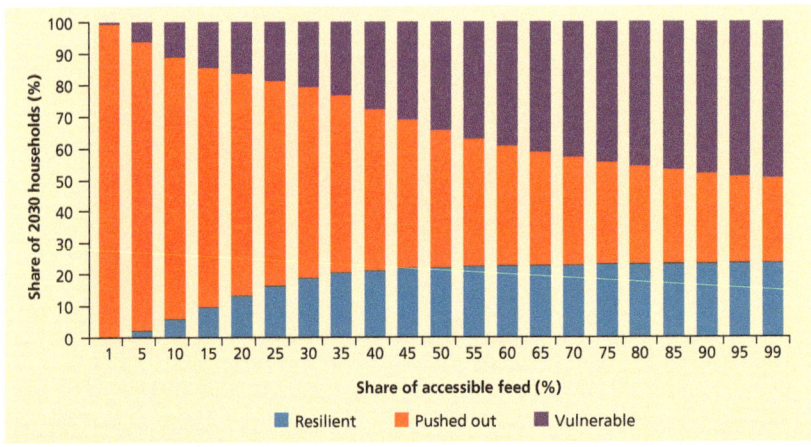

*Source:* De Haan 2016.

land use; (2) development of early warning and response systems to support early destocking when a drought shock is imminent and animals can be sold before they suffer a loss in value; and (3) programs that facilitate rapid restocking after the shock has passed. Experience suggests that such mechanisms can be both effective and efficient (Feinstein International 2007).

## Reducing sensitivity to shocks

Some pastoralists will be able to anticipate shocks and move their animals to avoid them, but others will be less fortunate and will be subjected to the full force of shocks when they occur. Those adversely affected by shocks are likely to include as well the many sedentary livestock-keepers whose reliance on farming activities keeps them anchored to particular locations.

Livestock-keepers living in dryland regions who are unable to move out of harm's way when a shock occurs will be affected only to the extent that their livelihood strategy is sensitive to the effects of the shock. For this reason, interventions that reduce sensitivity to shocks have the potential to significantly improve the performance of livestock systems in the drylands. Such interventions include: (1) improving preventive and clinical animal health services to protect livestock against infectious diseases and parasites; (2) developing infrastructure and funding to promote early offtake of male animals (young bulls), to be fattened in the higher-potential areas (highlands of East Africa and more humid areas of West Africa); and (3) promoting livelihood diversification among livestock-keeping households so that they can rely on alternative sources of income when the livestock enterprise fails.

The umbrella model was used to project the impact on the resilience of livestock-dependent households by 2030 of (1) improved animal health, and (2) early offtake of young male cattle (figure 5.4). The gains from these two interventions are relatively small when expressed as a proportion of all livestock-dependent households: the proportion of pastoral households owning enough TLU to be resilient would increase from 12 to 16 percent, and the number of agro-pastoral households having enough TLU to be resilient would increase from 20 to 32 percent. Still, the gains are significant when expressed in absolute terms: about 200,000 pastoral households and more than 3 million agro-pastoral households would become resilient by 2030, relative to the baseline. Similar numbers of households would emerge from the "non-viable" category, meaning they would no longer feel pressure to give up livestock-keeping. Interestingly, the projected benefits of these two interventions stand up under a range of weather scenarios.

An interesting—and unexpected—finding of the umbrella modeling exercise is that strengthening animal health services in the absence of complementary measures to increase feed supplies could lead to negative outcomes. Strengthening animal health services can accelerate growth rates, creating an opportunity to boost productivity and production, but accelerated growth rates in turn will increase feed requirements, putting further strain on what will already be a constraining factor (figure 5.5). Therefore, improvements in the delivery of animal health services will have to be accompanied by measures designed to make additional feed resources available, such as opening up under-exploited grazing areas or strengthening feed supply systems (figure 5.5).

**Figure 5.4** Impact of improved animal health and early offtake of young bulls on the resilience status of livestock-dependent households in 2030 (%)

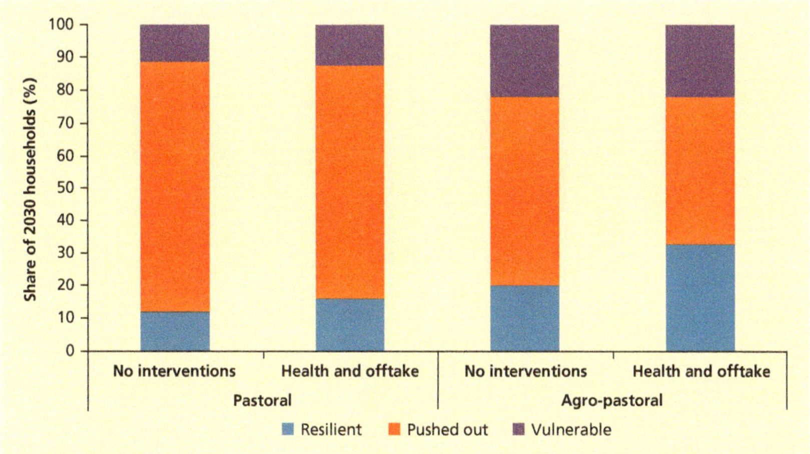

Source: De Haan 2016.

**Figure 5.5** Effect of weather on the effectiveness of improved animal health and early offtake of young bulls in improving the resilience of livestock-dependent households in 2030 (%)

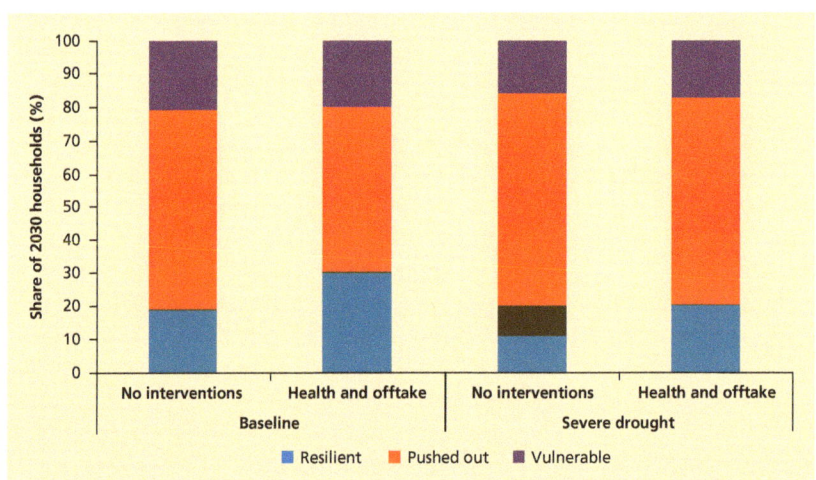

*Source:* De Haan 2016.

Figure 5.6 shows the projected impact by 2030 of improved animal health and early offtake of young male cattle on productivity and production. If implemented systematically throughout the drylands, these two practices would

**Figure 5.6** Average annual inputs and outputs for the different intervention scenarios compared to the baseline (%)

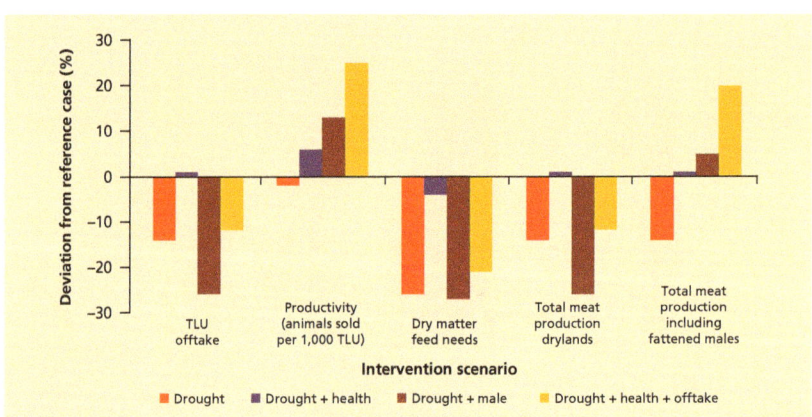

*Source:* De Haan 2016.

*Note:* The figures in the chart refer to the deviations from a reference scenario in which herd dynamics are driven by the same weather patterns observed in the period 1998–2011 and no policy intervention is in place.

increase offtake by about 25 percent and production of red meat by about 20 percent, resulting in an additional 750,000 MT (metric tons) of red meat produced annually by 2030. Feed requirements in the drylands would be reduced, although they would increase significantly in the more humid areas where increased fattening of cattle would occur.

Finally, early offtake of young male cattle would have a measurable impact on greenhouse gas emissions (figure 5.7).

**Figure 5.7** Greenhouse gas emissions for different interventions and climate scenarios in the two dryland study regions (kg)

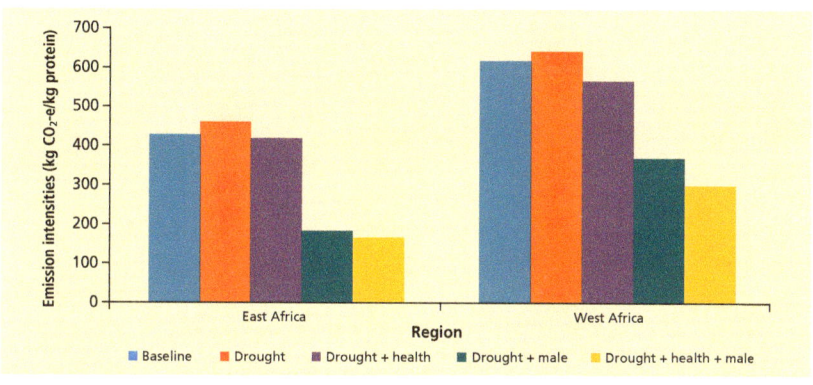

*Source:* De Haan 2016.

*Note:* Average cattle emission intensities (kg $CO_2$-e/kg protein), including males fattened in humid zones.

## Improving coping capacity

Livestock-keeping households in dryland regions—unable to move out of harm's way when shocks occur and having livelihoods that are sensitive to shocks—suffer frequent income losses. For these households the ability to survive will depend mainly on their coping capacity, that is, on their ability to draw on their own accumulated resources or resources provided by others to meet their needs during a critical period until their livelihood strategies can be reestablished.

Experience suggests that many livestock-keeping households, when hit by a shock, soon exhaust their limited accumulated resources, leaving them critically dependent on public programs. Public policy thus plays an important role in supporting the recovery process, particularly for non-resilient households. In considering the instruments available to the government, it is useful to distinguish between interventions that can be implemented relatively quickly versus interventions that require time to produce results.

Public interventions that can be implemented in the short run to strengthen the coping capacity of livestock dependent populations include (1) introducing insurance to provide compensation for lost animals and (2) establishing scalable safety nets to provide alternative sources of income until the livestock enterprise can be fully restored. (Scalable safety nets are discussed in detail in Chapter 9.)

Over the longer term, the objective of public policy should be to make the livestock-keeping population independent of outside support as much as possible. Given finite feed resources, the only way to increase significantly the number of resilient livestock-keeping households will be to address the current highly inequitable distribution of livestock assets.

The umbrella model was used to assess the likely impact of maintaining constant at current (2010) levels the grazing area available to households that are already resilient and allocating the remaining grazing area to vulnerable households, but in a consolidated manner that ensures that every vulnerable household gains access to a grazing area that is large enough to support enough TLU to ensure that the household is resilient (figure 5.8).

Directly allocating land and water access rights to vulnerable households while excluding resilient households, many of which own large herds, would obviously be challenging. It would not only come up against established distributions of political and economic power, but it would also run counter to the open access user rights systems that still prevail throughout most of the drylands. Still, it is possible to conceive of policies that could promote consolidation of grazing resources and lead to a more equitable redistribution, described as follows:

**Figure 5.8** Impact of consolidation of grazing area on the resilience status of livestock-keeping households, 2030

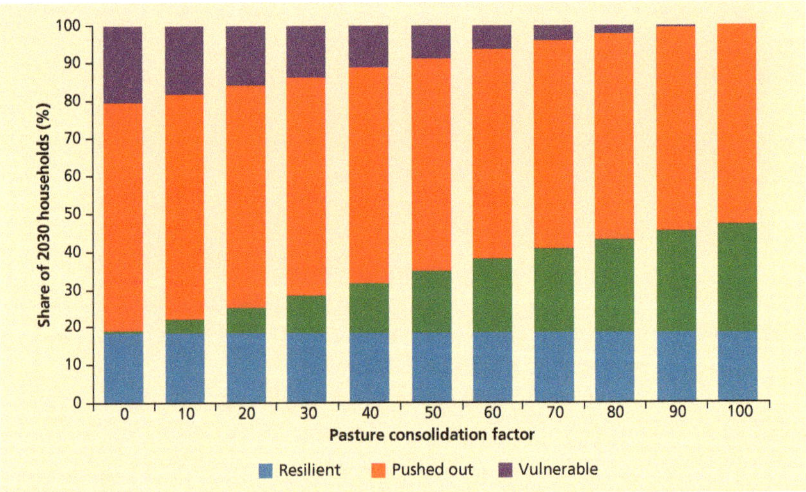

*Source:* De Haan 2016.

- Policies that limit land ownership (to prevent land grabbing by owners of large herds)
- Policies that enhance mobility of animals (to allow vulnerable households easier access to underutilized grazing resources)
- Policies that allocate exclusive water use and grazing rights for the wet and dry seasons to groups of smallholder livestock-keepers (to prevent denial of access by owners of large herds)

The second intervention—redistributing assets to allow less wealthy households to accumulate larger numbers of livestock—was modeled by estimating the impact of a change in the Gini coefficient (used as a proxy for the distribution of assets). A 50 percent increase in the Gini coefficient relative to the 2010 level would cut by one-half the number of vulnerable households likely to face pressure to exit from the sector (figure 5.9). Redistribution of assets, while always politically challenging, could in theory be achieved through the introduction of variable user fees or progressive tax policies, or both. At the practical level, a greater focus on the improvement of small ruminant production would also improve the distribution of livestock assets, as small ruminants are the main source of income for the poor.

None of the interventions described above, if introduced individually, would be expected to have a transformational impact on the numbers of vulnerable

**Figure 5.9** Impact of redistribution of assets on the resilience status of livestock-keeping households, 2030 (%)

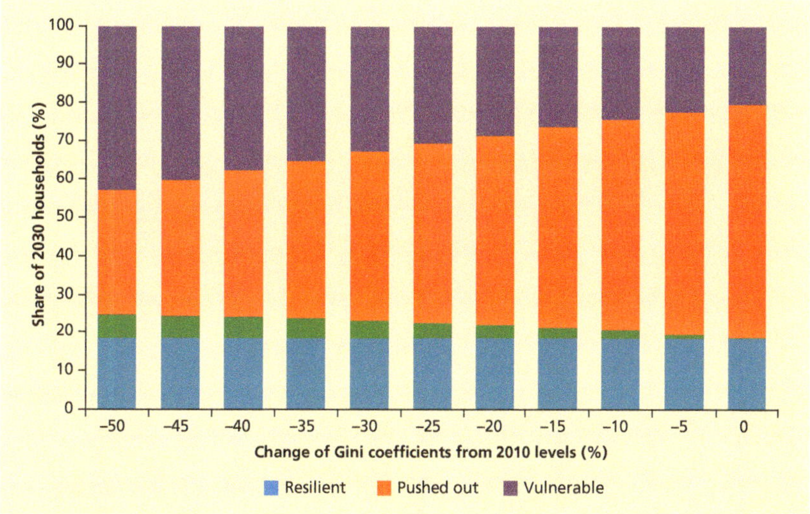

Source: De Haan 2016.

households. For this reason, the umbrella model was used to explore the combined impact of all the interventions. Combined, the interventions could make a difference: by 2030, the number of vulnerable households could be reduced to 16 percent, and the proportion of livestock-keeping households having so few animals that they would feel pressure to exit from the sector would be reduced to only 7 percent (figure 5.10).

**Figure 5.10** Impact of a combination of interventions on the resilience status of livestock-keeping households, 2030 (%)

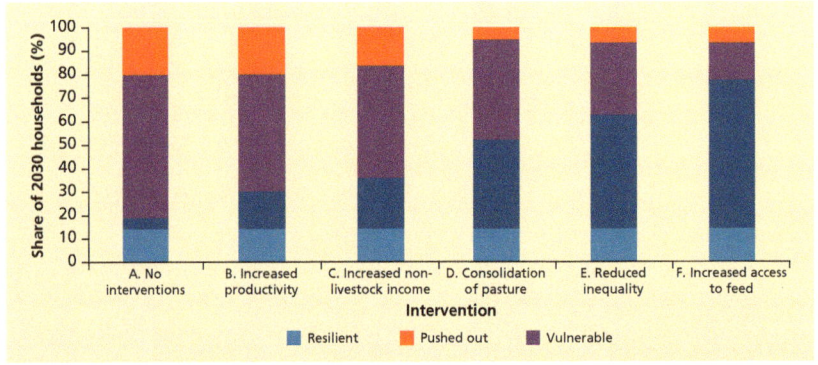

*Source:* De Haan 2016.

*Note:* Each intervention includes the effects of the ones preceding it; so, for example, intervention B includes the effects of intervention A; intervention C includes the effects of A and B; and so forth.

## Challenges

What are the obstacles to implementing these best-bet interventions designed to improve resilience among livestock-keeping populations in the drylands?

### Cost of increasing resilience

The first and perhaps most obvious challenge to overcome is cost. Analysis carried out for this book suggests that the unit cost of increasing resilience using the least-cost combination of interventions (that is, the unit cost of making one person or one household resilient) is relatively low, ranging from US$12/person/year to US$386/person/year, with an average US$27/person/year for all countries and systems (figure 5.11). Not surprisingly, the unit cost of providing resilience varies by country, by aridity zone, and by livestock system, and is significantly higher for pastoralists than for agro-pastoralists.

Using conservative assumptions, it is estimated that delivering improved animal health services and facilitating the early offtake of young male cattle would cost about US$0.5 billion per year for all the drylands of East and

**Figure 5.11** Cost effectiveness of health improvements and early offtake measures in improving the resilience status of households (US$)

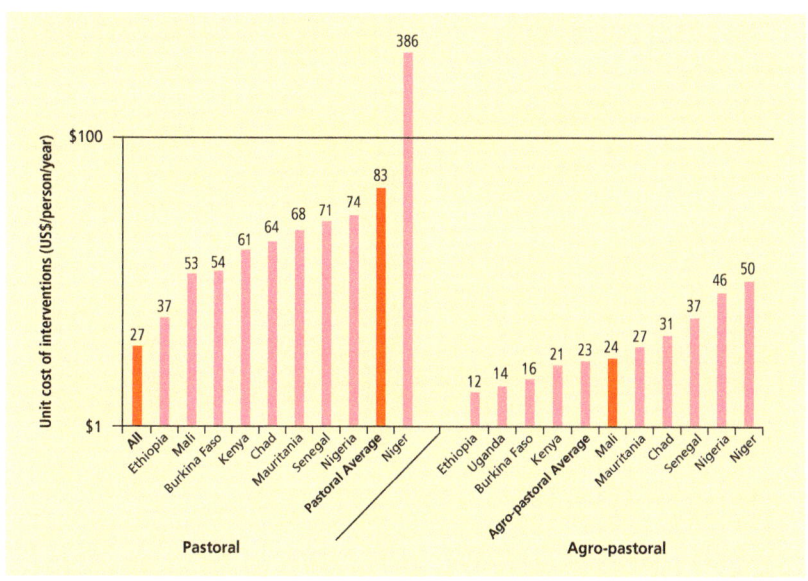

*Source:* De Haan 2016.

West Africa. While this amount is not insignificant, it is certainly smaller than the average value of the economic losses caused every year by droughts, disease outbreaks, civil conflict, and other shocks. It is also well below the cost of food aid, which currently averages US$4 billion/year in the Sahel and the Horn of Africa. Compared to the cost of providing humanitarian assistance when a shock has occurred, these interventions seem like an attractive option. While certainly not insignificant, an investment of about US$0.5 billion/year would likely yield a reduction of up to US$2 billion/year in humanitarian aid.

Mobilizing the necessary funding to support these interventions will be politically challenging, of course. The interventions require recurrent funding, which may prove difficult for many governments to mobilize. Perhaps development partners could be persuaded to help ensure that the necessary financial support can be sustained over the longer term (even permanently) by recognizing the savings that will be achieved in terms of reduced need for emergency assistance.

Aside from the overall cost, successful implementation of each intervention is associated with specific challenges—technical, economic, and institutional, including those associated with the management of common property resources (box 5.2).

**BOX 5.2**

## The challenge of managing common-pool resources in drylands

Most of the pastoralists in the drylands of East and West Africa share a strong ethos of open access to common-pool grazing resources. They believe that every pastoralist has the same rights to use grazing lands, regardless of ethnicity, nationality, seniority, or socioeconomic status. They emphatically argue that access is free and open for everyone; it does not matter where pastoralists come from, whether they are newcomers or old-timers or what is their ethnicity or nationality. For pastoralists, keeping cattle is not only a way of making a living, but also what makes life as pastoralists possible. In this sense, to deny cattle access to grazing resources is to deny pastoralists life (Moritz et al. 2013).

A large proportion of the rangelands that dominate Africa's drylands are open access. Historically there have been relatively few conflicts among African pastoralists over rights to common-pool grazing resources. Pastoralists do not live in a world made up only of pastoralists, however. They co-exist with other user groups, including farmers and fishermen, who do not share their ethos and practice of open access. Many farmers view grazing lands as lands that have not yet been made productive, and because often they do not recognize common property regimes and feel parcels can be appropriated for exclusive use by individuals, this constitutes a threat to common-pool grazing resources (Sayre et al. 2013). The result is agricultural expansion onto seasonal grazing lands and the transhumance corridors connecting them (Galvin 2009; Moritz 2006).

Many governments in East and West Africa have tried to protect pastoral resources and the rights of pastoralists to use these resources from agricultural expansion by designating agricultural and pastoral zones and delimiting transhumance corridors. These solutions have been implemented at local as well as national levels in the forms of rural or pastoral codes (Hesse and Trench 2000).

While much attention has been focused on problems of implementation and governance of rural codes (Flintan 2012; Hesse and Trench 2000; Tielkes and Schlecht 2001), there has been less discussion of the conflict between the flexibility and openness of the pastoral system and the fixing and delimitation of resources and resource use through the delimitation of pastoral zones and transhumance corridors. Turner (1999) has warned that there is a risk in formalizing pastoral tenure institutions into rural codes where flexibility is more appropriate for managing access to common-pool grazing resources, especially where there is considerable variation in the distribution of these resources through time and space. If tenure institutions become more formal and rigid, this can limit mobility, with potentially negative consequences for resilience.

*(continued next page)*

**Box 5.2** *(continued)*

Governments in East and West Africa have not always supported mobile pastoralists' use of common-pool grazing resources, for several reasons. First, while pastoralists are integrated into regional, national, and international livestock markets that reach millions of consumers, most of the trade is informal and invisible (Catley, Lind, and Scoones 2012; McPeak, Little, and Doss 2012). Governments therefore naturally favor the interests of agriculturalists whose production is more visible and more easily taxed (Behnke and Kerven 2013). Second, national laws are generally better at protecting the user rights of sedentary farmers over the grazing rights of mobile pastoralists, in part because mobile pastoralists do not remain in one location throughout the year, but also because pastoralists are not seen as making investments in the land, which is often a condition for obtaining tenure rights. Third, the processes of decentralization across Africa have resulted in more local control over natural resources, mostly at the level of municipalities. While decentralization works well for farmers who stay within a particular municipality throughout the year, that is not the case for mobile pastoralists who move through and use common-pool grazing resources in multiple municipalities over the course of a year. This means that decentralization and local control over natural resources are not accommodating mobile pastoral systems and are not appropriate for the governance of common-pool grazing resources in these systems (Turner 1999).

One of the key lessons of the "paradox of pastoral land tenure" is the need of pastoralists to secure access to pasture and water, but also to retain flexibility in resource use (Fernández-Giménez 2002). The critical lesson here is that governance needs to focus on supporting the flexibility of pastoral mobility in an open system, and this is not achieved by mapping, fixing, and delimiting the corridors, which may even have the opposite effect. The interests in support of pastoral mobility at the national and regional level are often not aligned with those at the local level, where government officials and traditional authorities tend to have primarily agricultural constituencies. At the national level, authorities benefit from the free movement of cattle because of taxes and other levies on pastoralists and livestock traders, whereas at the local level, authorities derive most of their income from agricultural populations.

*Source:* Adapted from Kerven and Behnke 2014.

## Improving animal health services

In the absence of private service providers, governments supported by development partners have often financed public provision of animal health services. Such efforts can be beneficial in the short run, but they usually prove counterproductive in the long run, as they undermine the incentives for private service providers to enter into the market once effective demand emerges. The challenge for policy makers is to create an incentive framework that can attract private service providers to enter into the market as public service providers are gradually phased out.

### Improving access to feed resources

Despite recent advances in legislation dealing with the pastoral economy, especially in the Sahelian countries, pastoral mobility is increasingly being hampered by the expansion of cultivated cropland. Land use rights in pastoral zones remain generally precarious, as often they are not recognized by institutions, especially in the strategic areas of lowlands, riverbanks, wet valleys, and forestry and pastoral reserves (Ickowicz et al. 2012). Policy reforms designed to formalize access by pastoralists to rangelands, coupled with investments in water resource development (to open up underutilized zones) and protection of corridors (to facilitate movement of animals to underutilized feed resources), could allow more complete use of available feed resources.

### Consolidating herd size and feed resources

Because of the highly inequitable distribution of livestock assets and the limitations on animal and feed resources, large numbers of households will not be able to accumulate the numbers of animals needed to generate enough income for them to remain above the poverty line. One way to overcome this problem would be to provide poor livestock-keepers with alternative sources of income, which would enable many of them to exit from the sector, freeing up resources for access by others. Facilitating exit from the sector—which is already occurring and will have to accelerate in future—is likely to be challenging from a policy perspective, but it represents an opportunity for poor households to transition into more productive and more sustainable livelihoods.

### Achieving more equitable distribution of livestock resources

Evidence is accumulating that livestock ownership both in the Horn of Africa and in the Sahel is becoming increasingly concentrated. Ever greater numbers of animals are ending up in the hands of wealthy traders and government officials, who tend to manage their herds using hired labor, which crowds out many of the small-scale herders who make up by far the largest share of the livestock-keeping population. If this trend could be reversed, the households able to accumulate the numbers of animals needed to stay above the poverty line could increase significantly. Progressive taxation of livestock assets and imposition of user fees in public rangelands could discourage accumulation of large herds, but such policies are likely to engender significant resistance from politically and economically influential livestock owners.

## Key messages

The analysis summarized here makes clear that there is scope for expanding livestock production in drylands and increasing the contribution of drylands producers to the rising demand in Sub-Saharan Africa for animal-source products. Policy changes and supporting investments such as those described here could halve the regional deficit projected to emerge by 2030.

The results of the modeling exercise suggest that feed and animal resources will be insufficient to provide secure and adequate livelihoods for all of the people in the drylands who depend on livestock as their principal livelihood source. Under the BAU scenario, by 2030 about 77 percent of pastoralist households and 58 percent of agro-pastoralist households will not be able to accumulate the numbers of animals needed to generate enough income for them to subsist even at 50 percent of the poverty line. The current inequitable distribution of livestock assets, which is projected to become worse as a result of the ongoing transformation of the dryland economy, is likely to put further pressure on poor pastoralists.

Fortunately, these gloomy scenarios can be avoided. Investments in improving animal health services and increasing market integration, combined with measures to improve access to the available feed resources, could increase the share of livestock-keeping households able to accumulate enough animals to remain resilient. Adoption of the full package of best-bet interventions could reduce the share of livestock-keeping households who feel pressure to exit from the sector to as little as 7 percent.

The development of alternative sources of income, inside or outside the drylands, needs to be an integral and major component of any dryland development strategy. Going forward, the traditional narrow focus on increasing production of milk and meat will have to change so as to embrace a wider range of diversified income-generating activities. There is need as well to strengthen the incentives for livestock-keepers to serve as responsible stewards of the environment.

Government policies designed to sedentarize pastoralists, particularly in the more arid zones, are unlikely to succeed. Herds and flocks must be mobile if they are to use temporally and geographically distributed feed resources, so measures that restrict their mobility will reduce productivity and exacerbate poverty.

## Notes

1.  The Tropical Livestock Unit (TLU) is an artificial construct that can be used to aggregate different livestock species. For Sub-Saharan Africa, the conversion factors are: 1 camel = 0.7 TLU, I cow = 0.6 TLU, and 1 sheep or goat = 0.1 TLU.
2.  Resilient households are defined as households owning at least the minimum number of TLU needed to stay above the poverty line, assuming that 70 percent of the income of pastoralists is derived from livestock, and 35 percent of the income of agro-pastoralists.Three categories are distinguished: (a) resilient households = households owning more than 15 TLU, (b) vulnerable households = households owning 7.5 to 15 TLU, and (c) non-viable households = households owning less than 7.5 TLU and likely to be forced to seek an alternative livelihood strategy. These levels increase with drought and decrease with the introduction of productivity-enhancing innovations. For details, see De Haan (2016).

## References

Behnke, R.H., and C. Kerven. 2013. "Counting the Costs: Replacing Pastoralism with Irrigated Agriculture in the Awash Valley, North-Eastern Ethiopia." Climate Change Working Paper No. 4. International Institute for Environment and Development, London.

Breman, H., and C.T. de Wit. 1983. "Rangeland Productivity and Exploitation in the Sahel." *Science* 221: 1341–47.

Catley, A., J. Lind, and I. Scoones, eds. 2012. *Pastoralism and Development in Africa: Dynamic Changes at the Margins.* London: Routledge (Earthscan).

De Haan, C., E. Dubern, B. Garancher, and C. Quintero. 2014. "Pastoralism Development in the Sahel: A Road to Stability?" World Bank Global Center on Conflict, Security, and Development, Nairobi.

De Haan, C., ed. 2016. *Improved Crop Productivity for Africa's Drylands.* World Bank Studies. Washington, DC: World Bank.

Derrick, J. 1977. "The Great West African Drought, 1972–74." *African Affairs* 76: 537–86.

Desta, S., and D.L. Coppock. 2004. "Pastoralism Under Pressure: Tracking System Change in Southern Ethiopia." *Human Ecology* 32(4): 465–86.

Feinstein International Center. 2007. "Impact Assessments of Livelihoods-based Drought Interventions in Moyale and Dire Woredas. A Pastoralist Livelihoods Initiative Report." Feinstein International Center, Medford, MA.

Fernández-Giménez, M.E. 2002. "Spatial and Social Boundaries and the Paradox of Pastoral Land Tenure: A Case Study From Postsocialist Mongolia." *Human Ecology* 30(1): 49–78.

Flintan, F. 2012. "Making Rangelands Secure: Past Experience and Future Options." International Land Coalition, Rome.

Galvin, K. A. 2009. "Transitions: Pastoralists Living with Change." *Annual Review of Anthropology* 38: 185–198.

Gerber, P.J., H. Steinfeld, B. Henderson, A. Mottet, C. Opio, J. Dijkman, A. Falcucci, and G. Tempio. 2013. *Tackling Climate Change through Livestock: A Global Assessment of Emissions and Mitigation Opportunities.* Rome: Food and Agriculture Organization of the United Nations (FAO).

Ham, F., and E. Fillol. 2012. "Pastoral Surveillance System and Feed Inventory in the Sahel." In Michael B. Coughenour and Harinder P. S. Makkar, editors, *Conducting National Feed Assessments: FAO Animal Production and Health Manual No. 15,* 83–94. Rome, Italy: FAO.

Headey, D., A.S. Taffesse, and L. You, 2014. "Diversification and Development in Pastoralist Ethiopia." *World Development* 56: 200–213.

Hesse, C., and P. Trench. 2000. "Who's Managing the Commons? Inclusive Management for a Sustainable Future." *Securing the Commons (1).* SOS Sahel International (UK) and the Drylands Programme of the International Institute for Environment and Development (IIED).

Ickowicz, A., V. Ancey, C. Corniaux, G. Duteurtre, R. Poccard-Chappuis, I. Touré, E. Vall and A. Wane. 2012. "Crop-livestock Production Systems in the Sahel: Increasing Resilience for Adaptation to Climate Change and Preserving Food Security." In A. Meybeck, J. Lankoski, S. Redfern, N. Azzu, and V. Gitz, *Building Resilience for Adaptation to Climate Change in the Agriculture Sector: Proceedings of a Joint FAO/OECD Workshop 23–24 April 2012.* Rome: FAO.

Kerven, C., and R. Behnke, eds. 2014. "Human, Social, Political Dimensions of Resilience." Unpublished paper, FAO, Rome.

Lesnoff M., C. Corniaux, and P. Hiernaux. 2012. "Sensitivity Analysis of the Recovery Dynamics of a Cattle Population Following Drought in Sahel." *Ecological Modeling* 232: 28–39.

McPeak, J., P.D. Little, and C. Doss. 2012. *Risk and Change in an African Rural Economy: Livelihoods in Pastoralist Communities.* Routledge ISS Studies in Rural Livelihoods (Book 7). The Hague: Routledge.

Moritz, M. 2006. "Changing Contexts and Dynamics of Farmer-Herder Conflicts across West Africa." *Canadian Journal of African Studies* 40: 1–40.

Moritz, M., P. Scholte, I.M. Hamilton, and S. Kari. 2013. "Open Access, Open Systems: Pastoral Management of Common-Pool Resources in the Chad Basin." *Human Ecology* 41(3): 351–365.

Niamir-Fuller, M., ed. 1999. *Managing Mobility in African Rangelands: The Legitimization of Transhumance.* London: Intermediate Technology.

Oxfam. 2012. "Food Crisis in the Sahel: Five Steps to Break the Hunger Cycle in 2012." Joint Agency Issue Briefing. May 31. Oxfam, Oxford, United Kingdom.

Place, F., and J. Binam. 2013. "Economic Impacts of Farmer-Managed Natural Regeneration in the Sahel." End-of-Project Technical Report for Free University Amsterdam and IFAD, World Agroforestry Center, Nairobi, Kenya.

Pratt D.J., F. LeGall, and C. De Haan. 1997. *Investing in Pastoralism: Sustainable Natural Resource Use in Arid Africa and the Middle East.* Washington, DC: World Bank.

Reij, C., G. Tappan, and M. Smale. 2009. "Agroenvironmental Transformation in the Sahel: Another Kind of 'Green Revolution'." Discussion Paper 00914, International Food Policy Research Institute (IFPRI), Washington, DC.

Robinson, J., and F. Pozzi. 2011. "Mapping Supply and Demand for Animal-Source Foods to 2030." Animal Production and Health Working Paper. No. 2. FAO, Rome.

Robinson, T.P., G.R.W. Wint, G. Conchedda, T.P. Van Boeckel, V. Ercoli, E. Palamara, G. Cinardi, L. D'Aietti, S.I. Hay, and M. Gilbert. 2014. "Mapping the Global Distribution of Livestock." PLoS ONE 9(5): e96084.

Sayre, N., R. McAllister, B. Bestelmeyer, M. Moritz, M. Turner. 2013. "Earth Stewardship of Rangelands: Coping with Ecological, Economic and Political Marginality." Frontiers in Ecology 11 (7): 348–54.

SIPSA, 2012 Atlas of Trends in pastoral systems in the Sahel 1970-2012. FAO-CIRAD.

Summers, L.H. 1992. "Investing in All the People." World Bank Policy Research Working Papers. World Bank, Washington DC.

Tielkes, E., and E. Schlecht. 2001. "Elevage et gestion de parcours au Sahel, implications pour le développement—Livestock production and range management in the Sahel, implications for development." Comptes rendus d'un atelier régional ouest-africain sur le thème "La gestion des pâturages et les projets de développement: quelles perspectives?" tenu du 2 au 6 octobre 2000 à Niamey, Niger.

Turner, M.D. 1999. "Conflict, Environmental Change, and Social Institutions in Dryland Africa: Limitations of the Community Resource Management Approach." Society and Natural Resources 12(7): 643–57.

UNESCO (United Nations Educational, Scientific, and Cultural Organization). 2010. "Education Counts: Towards the Millennium Development Goals." UNESCO, Paris. http://unesdoc.unesco.org/images/0019/001902/190214e.pdf

Wint, G.R.W., and T.P. Robinson. 2007. "Gridded Livestock of the World." FAO, Rome.

## Chapter 6

# Tree-Based Systems: Multiple Pathways to Boosting Resilience

*Frank Place, Dennis Garrity, Paola Agostini*

## Current situation

Tree-based production systems have enormous potential to reduce vulnerability and increase the resilience of households in dryland regions of Sub-Saharan Africa. Trees are key providers of biomass, which is critical for many livelihood needs. Wood from trees is the leading source of energy in many dryland countries and is an important construction material. Foliage and pods from trees and shrubs are the most important source of feed for camels and goats, the dominant livestock species in more arid parts of the drylands. Trees and shrubs offer enhanced sources of the organic matter needed to improve the structure and raise the fertility of soils used for agriculture. In addition, many parts of trees provide different medicinal products for people. And fruits and vegetable foliage harvested from trees are important seasonal food sources for people living in drylands and for sale.

The benefits from trees take on added value when it is considered that tree-based production systems are relatively impervious to many of the shocks that affect other production systems, especially livestock-keeping and agriculture. With their deep roots, trees maintain their standing value and offer some production even in drought years. Therefore they are a good buffer against climatic risk and a critical element in a diversification strategy designed to maintain levels of consumption and income in good times and bad. In addition, their value can be tapped when it is most needed: wood from trees can be harvested throughout the year, and many annual tree products are harvested at times different from the times when annual crops are harvested.

The term "tree-based systems" as used in this book refers to agricultural systems, forest/woodland/bushland systems, or pastoral (rangeland) systems in which trees play a significant role. Within each of these three main classes of land use, many different tree species can be economically and ecologically important.

## Management strategies for tree-based systems

Not surprisingly considering their variability, tree-based systems encompass a wide range of management practices. It is important to distinguish between tree-based systems that involve the *managed regeneration* of trees (often indigenous species) and tree-based systems that involve *purposeful planting and/or management* of trees (often introduced species).

### Natural regeneration

Managed regeneration of indigenous tree species can lead to the emergence of diversified tree-based systems capable of generating multiple products and services. In the drier areas of Sub-Saharan Africa regeneration accounts for a large majority of the trees being managed by farmers. Regenerative practices include farmer-managed natural regeneration (FMNR) of trees found in croplands, as well as assisted natural regeneration (ANR) involving the use of enclosures to rehabilitate rangelands or woodlands. Systems based on natural regeneration typically include a diverse set of tree species that are well-adapted to local conditions and that entail relatively low establishment costs. Regenerative systems are currently being expanded in large areas throughout the arid and semi-arid drylands. Regeneration of trees on farms occurs throughout the farm, including on crop fields. The result is a mosaic of trees integrated into other land uses such as cropping, pastures, and fallows.

FMNR on agricultural lands and ANR on community lands represent cost-effective ways of achieving widespread increases in the numbers of valuable, adapted, and diverse trees. What these practices have in common is that in both cases, people (individual farmers or entire communities) actively influence natural biological regeneration processes to achieve patterns that better suit their needs. On agricultural lands, farmers identify naturally regenerating tree seedlings in their fields and manage them to provide various benefits (for direct products and for crops or livestock). On community lands, community groups may adopt the same practices, and they may also introduce grazing management systems at the community level designed to allow successful tree regeneration in targeted areas. Under both systems, protecting and weeding around young trees may be necessary to help them survive.

In recent years FMNR has gained in popularity in many dryland areas throughout Sub-Saharan Africa. Because FMNR requires minimal cash investment, it can expand rapidly through farmer-to-farmer and village-to-village diffusion. The more than 5 million hectares of medium- to high-density tree cover newly regenerated on croplands in Niger provide a dramatic example of how quickly and how extensively the practice can spread (Reij, Tappan, and Smale 2009). And Niger may be just the tip of the iceberg. A recent study carried out in Niger, Mali, Burkina Faso, and Senegal found that almost all farmers are actively regenerating trees (Place and Binam 2013).

The benefits derived from FMNR vary from location to location, depending on which tree species are present in the area and what products and services are valued locally. Throughout the Sahel more than 100 woody species are being managed by farmers through natural regeneration. These trees are of high value: they contribute products for human consumption (more than US$200 per household per year) and feed for livestock during the late dry season, and they have positive effects on crop yields (accounting for roughly 20–25 percent of variation in millet and sorghum yields).

## Purposeful planting

Purposeful planting and/or management of certain types of tree species that can produce economically valuable products and services are also important in the drylands, particularly in dry subhumid zones where rainfall is more plentiful. Where the water supply is more assured, the costs of planting trees are lower, the risk of losing trees to drought is less pronounced, and the productivity of trees is higher.

## Benefits of tree-based systems

Whether based on managed regeneration or purposeful planting, tree-based systems in drylands are capable of generating many economically valuable products and services.

### Improved soil fertility

Trees of all types have properties that are beneficial for soil fertility. These include root systems that hold soils in place, litter that falls as mulch, and organic matter that the roots and litter provide to micro and macro fauna in the soil. Many farmers have known and appreciated these properties for generations. At the same time, trees can compete with crops for nutrients, water, and light, so farmers must weigh the costs and benefits before associating trees with crops. The presence of trees in crop fields may also complicate plowing, which is why extension agents often convey messages about cultivating "clean" fields (Smith 2010).

Quite a number of tree species have been found to offer significant soil fertility benefits in dryland regions of Africa. Unquestionably the most important of these is *Faidherbia albida* (formerly *Acacia albida*), which fixes nitrogen from the atmosphere, develops a deep rooting system that allows it to access underground moisture during times of drought, produces a light canopy that does not compete much with underlying crops, and drops its nitrogen-rich leaves in advance of the rainy season. Many other species similarly contribute to improved soil fertility, for example, many of the acacia species.

In drier zones characterized by less than 600 millimeters of annual rainfall, virtually all fertilizer trees are established through FMNR. In more humid reaches of the drylands, where population densities are generally higher and the incentives and capacities for intensification are higher, hundreds of thousands of farmers have been induced to establish fertilizer trees through purposeful planting (Garrity et al. 2010).

A meta-analysis of studies on the effects of fertilizer trees on maize yields found that such trees often have significant positive effects; even doubling of yields is not uncommon (Sileshi et al. 2008). The effects can be quite variable, however, with species choice, management practices, and environmental conditions all playing critical roles. Two recent studies examined the yield and profit effects from FMNR of *Faidherbia*-based systems in Malawi (Glenn 2012) and the Sahel (Place and Binam 2013). Both studies found that the trees had positive effects on yields and profits. In multiple locations in Mali, Burkina Faso, and Niger, *Faidherbia* and other species established through FMNR boosted yields of millet and sorghum from 16–30 percent, controlling for other inputs (Place and Binam 2013). In multiple locations in Malawi, *Faidherbia* trees boosted maize yields by 12–16 percent, also controlling for other inputs (Glenn 2012). In addition to helping increase yields during times of normal rainfall, fertilizer trees provide some protection against drought. The available evidence, while limited, suggests that yield decreases are generally less pronounced during droughts when fertilizer trees are present in the field (Akinnifesi et al. 2010).

In the more humid parts of the drylands the benefits of fertilizer trees can be realized rapidly, especially in planted systems, because planted trees quickly produce large quantities of biomass containing significant amounts of nitrogen (more than 100 kilograms of nitrogen per hectare [kg N/ha]). In the drier parts of the drylands the benefits of fertilizer trees take longer to appear because the trees that make up the mainly regenerative systems that dominate in the drier zones take longer to become established. In addition to contributing to improved soil fertility through the production of leaf biomass, trees can help to build up soil biological and physical health through the continual deposition of organic matter. Organic matter improves the resilience of the soil resource, so that it is more productive for a wider range of crops and other plants. The positive effects of trees on soil carbon (e.g., Beedy et al. 2014; Nair et al. 2009), soil water retention capacity (Mafongoya et al. 2006), and soil fauna (Mafongoya, Kuntashula, and Sileshi 2006) are supported by a large body of evidence.

Case studies have shown that both regenerative and planted tree systems can be profitable (for examples see Ajayi et al. 2007; Ajayi et al. 2011; Place and

Binam 2013). Planted systems require more labor, not only for establishment of trees but also for management of potential competition with crops, especially when exotic fast-growing tree species are used. The added labor costs are more than compensated, however, through higher crop yields. Ajayi et al. (2007) found that the net present value from a five-year improved fallow rotation (two years fallow followed by three years of maize) ranged from US$270–310 per hectare [ha], compared to US$130 per hectare for the conventional system with no fertilizer. Although systems based on the use of fertilizer outperform tree-based systems in terms of crop yield and net present value, the two systems are comparable in terms of benefit-cost ratio and returns to labor.

## Livestock fodder and feed

Trees and shrubs produce feed for livestock, particularly during the dry season when natural pasture is scarce. For this reason, farmers use many dryland trees and shrubs to nourish their livestock. In West Africa, two of the most common are *Pterocarpus spp* and *Piliostigma spp*.

The limited available evidence on the effects of trees and shrubs on livestock growth in drylands comes mainly from researcher-managed feeding trials. For example, supplementation of pasture in Zimbabwe with 75 grams of *Acacia angustissima* fed to a group of goats each day was found to result in an incremental increase of 36 grams per goat per day (Mukandiwa et al. 2010).

Relatively little research has been done at the farm level to assess the profitability of tree investments in the livestock sector. Such assessments are complicated by the large number of tree species used for feed, the high level of variability in the duration and frequency of feeding, and the shifting composition of feed resources, among other factors. Place and Binam (2013) found positive correlations between the number of goats and sheep and the number of fodder shrubs on farms in Burkina Faso, but no such correlation was detected in neighboring countries. The same authors also found positive correlations between the value of goat and sheep production and the production value-to-stock ratio on the one hand and the number of trees on the other. This suggests that at least in the case of small ruminants, private investment in fodder trees and shrubs is associated with higher animal stocks and production.

## Fuel wood and timber

Trees are the leading source of energy in almost all rural areas of Africa, including the drylands. Firewood and charcoal are widely used for cooking, bathing, laundering, and heating. In many countries, the drylands are a major supplier of firewood and charcoal for urban areas. The value of traded charcoal is currently estimated to be in the billions of dollars, making charcoal one of the most valuable commodities traded in the region. Current fuel wood production comes mainly from off-farm sources, and harvesting methods are frequently destructive to the environment. Governance of fuel wood production and marketing is generally weak, which creates uncertainty throughout the value chain, gives rise to extra-legal transactions costs, and also undermines incentives for long-term investment. Reforms to policy and regulatory frameworks could significantly improve the management of fuel wood harvested from woodlands, as well as strengthen the incentives to source fuel wood from farms.

Tree products (especially timber and poles) are important construction materials in many dryland regions of Africa. Timber and pole production almost always involves the purposeful planting of seedlings, because the profitability depends critically on the use of quality germplasm and adoption of careful management practices. Timber and pole production therefore are best suited to areas in which rainfall is more abundant and more reliable, especially the dry subhumid zone. Timber and pole production schemes in Sub-Saharan Africa have for the most part relied on exotic species, such as *Eucalyptus camuldulensis* or *Acacia mearnsii*. In addition, many indigenous trees with high value have the potential to perform well, as long as sufficient attention is paid to germplasm selection and management. For example, *Melia volkensii* already supports a thriving high-quality furniture wood production industry in Kenya.

### Non-wood tree products

Trees and shrubs in the drylands produce many non-wood products that are extensively harvested for home consumption as well as for sale. These non-wood products include foods (fruits, nuts, and leaves); medicines; gums and resins; oils and fragrances; and fodder for livestock. The value of non-wood products varies considerably by region. Baobab contributes significantly to incomes in Senegal; shea in Burkina Faso, Mali, and northern Ghana; gum arabic in Sudan; and marula in southern Africa. Cashew is another important commodity, prominent in the semi-arid and subhumid zones. Over 1.5 million farmers grow cashews in Africa, and production doubled between 2003 and 2011. Fruit production, while still relatively limited, has tremendous potential, as fruit consumption is growing rapidly throughout the region as a result of urbanization and improved nutrition awareness. Production of many of these non-wood tree products can be expanded to meet growing export demand. In some cases the opportunities lie more with value addition than with production. For example, the fruits of hundreds of millions of shea trees are processed locally using traditional methods to meet domestic demand or are exported unprocessed. Investments in industrial processing machinery could significantly increase the quantity and quality of shea nut products, generating increased profits for producers, processors, and exporters, and boosting foreign exchange earnings for exporting countries. A similar situation prevails in the case of cashew.

### Environmental services

Trees provide many environmental services, including carbon sequestration, watershed protection, and soil health enrichment. All trees sequester carbon at a relatively stable proportion of 0.5 of the woody biomass dry weight. Tree growth is slower as aridity increases, and the annual aboveground carbon sequestration from a typical regenerated field may be around 1 ton per hectare in the semi-arid regions with an additional third of that below ground.

The value of trees and tree products can be significant, both in terms of the contribution to total household income, as well as in terms of cash income from sales (figure 6.1).

## Opportunities

How might the benefits produced through tree-based systems contribute to the resilience of households living in drylands? To answer that question, it is useful to consider the potential impacts of trees on the three determinants of resilience.

### Reducing exposure

There is some evidence that wide-scale adoption of tree-based systems can actually affect weather patterns in the drylands, for example by tempering the frequency and the strength of storms. These effects are at best very minor, however, and almost certainly below the level needed to significantly reduce exposure to shocks.

**Figure 6.1** Revenue from sales of tree products, selected countries, West Africa (%)

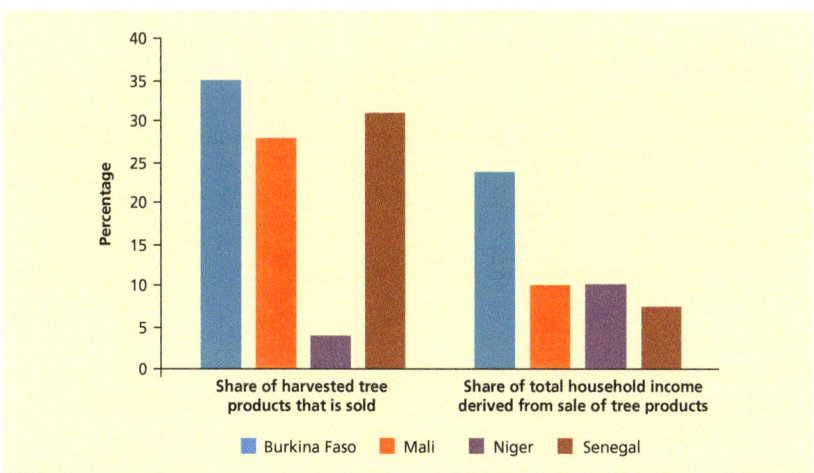

*Source:* Place et al. 2016.

### Reducing sensitivity

While trees may not reduce exposure to shocks, they can play an important role in reducing household sensitivity to shocks. Trees are not completely impervious to climate change, but their deep rooting systems allow them to take advantage of moisture stored in the soil, which makes them less vulnerable to seasonal rainfall reductions. This robustness allows trees to play a particularly important role in reducing sensitivity to at least two types of shocks experienced in the drylands: weather-related shocks and health-related shocks.

**Reduced sensitivity to weather-related shocks.** The dominant weather-related shock in the drylands is droughts that are severe, frequent, or prolonged. Trees growing in crop fields attenuate the severity of drought effects on crop performance by modifying the microclimate. Crops growing in the vicinity of trees experience a more favorable microclimate, with significantly higher humidity in the crop canopy causing a lower vapor pressure deficit. Trees can also lower solar radiation stress experienced by crops, and they can increase the infiltration and storage of rainfall in the soil by reducing surface runoff. The additional biomass that trees provide increases soil organic matter, which enhances soil moisture storage and improves nutrient availability to crops. Moreover, there are circumstances under which some trees effectively transfer water from deeper depths up to near the soil surface through their root systems and make such water available to nearby crops, a phenomenon known as "hydraulic lift" (Bayala et al. 2014). These various features of trees combine to reduce the rate of onset of crop water stress, enabling crops to more successfully withstand periods of drought during the growing season.

A second weather-related shock in the drylands is heat. All crops experience a reduction in yield whenever temperatures exceed a certain threshold level. High temperatures depress yields through two processes. First, plants respond to high temperatures by increasing their respiration rate, which causes them to burn up more energy, leaving less available for grain filling. Second, high temperatures shorten the crop maturity period, which reduces the size and weight of the grain. Trees growing in crop fields can significantly reduce temperatures in the crop canopy and soil, particularly during the middle part of the day. Across the growing season, avoiding daily temperature shocks can allow plants to photosynthesize longer, leading to increased grain filling and higher yield. These effects can be observed in the more stable crop yields recorded during drought years in fields containing trees than in fields without trees (for example, see evidence from Niger cited in Reij, Tappan, and Smale 2009). Survey data are consistent with testimony by many farmers that higher tree populations reduce drought effects.

**Reduced sensitivity to health-related shocks.** Trees can also help reduce sensitivity to health-related shocks. Fruits and vegetable foods harvested from trees comprise part of the regular diet in the drylands, and in many cases they are critical for good nutrition because they contain vitamins and micronutrients that are unavailable from other sources. For example, the fruits and leaves of baobab are highly nutritious in vitamins A and C, which are lacking in staple foods (Orwa et al. 2009). Tree-based foods take on special significance during periods of seasonal or prolonged drought-induced hunger when crops and animal-source foods become unavailable (Place and Binam 2013).

Crop modeling carried out for the Africa Drylands study and further discussed in Chapter 12 helped provide orders of magnitude of the benefits of FMNR in terms of reduction of drought impacts. When FMNR of native species is added to the other productivity-enhancing technologies discussed in this book, the effects are impressive. In a group of 10 countries in East and West Africa, the projected number of poor, drought-affected people living in drylands in 2030 falls—compared to the business as usual (BAU) no intervention scenario—by 13 percent with low-density tree systems and by more than 50 percent with high-density tree systems (figure 6.2).

## Improving coping capacity

In addition to reducing sensitivity to shocks, trees can enhance the capacity of households in drylands to cope with the effects of shocks after the shocks have occurred. Trees are assets that can be cut and sold for cash or exchanged for

**Figure 6.2** Estimated reduction in the average number of drought-affected people through use of FMNR and other technologies by year 2030 (millions)

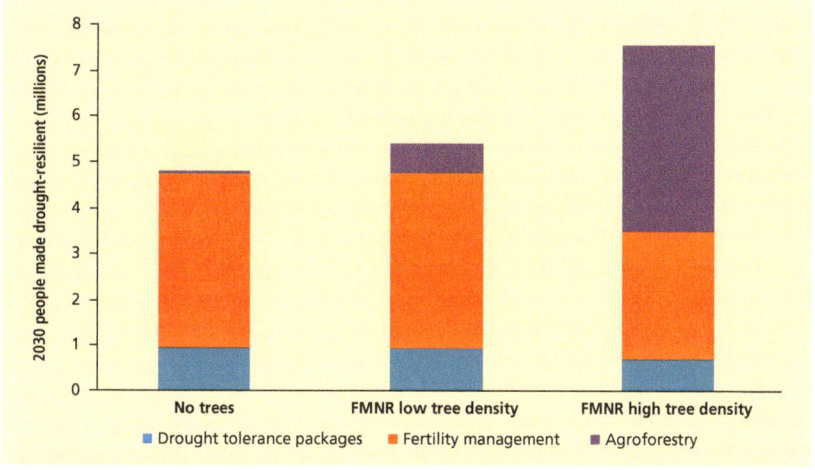

Source: Calculation based on the approach discussed in the Appendix.

Note: FMNR = farmer-managed natural regeneration. The data in the chart refer to the number of households that by 2030 could become resilient to droughts, on an annual average, by adopting different packages of resilience interventions. The figure presents aggregated results for Ethiopia, Kenya, Uganda, Senegal, Nigeria, Mauritania, Chad, Mali, Burkina Faso, and Niger.

goods in times of need. In the Maradi and Zinder Regions of Niger, where 1.2 million households now sustain medium to high densities of tree populations on their farms, farmers cut tree branches on a continuous cycle for household fuel wood supplies and for sale, and some mature trees are cut down and sold in local wood markets for poles and construction materials. Export markets are active in shipping wood south to Nigeria. During prolonged drought periods these tree assets may be gradually liquidated to supply the household with cash for food purchases. This process was observed to be an important source of coping capacity for households during recent droughts (Reij, Tappan, and Smale 2009).

## Returns to investment

The rapid expansion of FMNR throughout large areas of West and East Africa suggests that farmers in the drylands value the benefits of the technology. But just how profitable is the technology, especially in comparison to other technologies that farmers could choose to adopt? Researchers have been homing in on this question, although definitive answers remain elusive due to the difficulty of measuring all of the multiple benefits and the long periods over which they are realized.

Place et al. (2016) explored the returns to investment in FMNR using a model constructed to analyze costs and benefit streams over a 20-year period. The model can be calibrated to represent the situation prevailing in different aridity zones and in different countries; in this case, the focus is on parkland systems in Mali and Niger in which millet is the dominant crop. The investment considered is FMNR, starting from a base of no trees and allowing the tree density to build up to the average density observed in the two countries. Two benefit streams are captured: the value of direct tree products (wood and non-wood), and the value of improved crop yields. Three cost categories are considered: (1) establishment of the system, (2) annual costs (upkeep and harvesting) related to tree products, and (3) annual costs related to crop production. Discount rates of 10, 15, and 20 percent are used over 20- and 30-year time frames.

Table 6.1 shows the net present value (NPV) and benefit-cost ratio (BCR) for six combinations of discount rates and time periods (all other variables are fixed). The estimated returns range from a low of NPV = US$29.9 per hectare and BCR =1.5 (for the 20-year period assuming a 20 percent discount rate) to a high of NPV = US$178.11 per hectare and BCR = 2.66 (for a 30-year period assuming a 10 percent discount rate). The IRR (internal rate of return) does not vary much for the different assumptions. The IRR is 34 percent in a 20-year time frame and 36 percent in a 30-year time frame. The break-even year similarly does not vary much, falling in year 11 in the case of a 20 percent discount rate and in years 10 and 9 in the case of discount rates of 15 percent and 10 percent, respectively. The benefit streams per hectare from crops and tree products are

**Table 6.1** Private economic returns from FMNR (US$ per hectare)

| | Mali | | Niger | |
|---|---|---|---|---|
| | 20-year period | 30-year period | 20-year period | 30-year period |
| **Net present value (NPV)** | | | | |
| 10% discount rate | 133.57 | 178.11 | 442.80 | 568.99 |
| 15% discount rate | 66.82 | 82.46 | 253.94 | 298.24 |
| 20% discount rate | 29.89 | 35.71 | 149.18 | 165.66 |
| **Benefit-cost ratio (BCR)** | | | | |
| 10% discount rate | 2.43 | 2.66 | 6.78 | 7.47 |
| 15% discount rate | 1.94 | 2.09 | 5.40 | 5.83 |
| 20% discount rate | 1.52 | 1.60 | 4.19 | 4.43 |

*Source:* Place et al. 2016.

virtually the same in the case of Mali. In contrast, all the economic variables are more favorable for FMNR in Niger, due to larger benefit streams from both harvested tree products and crop yields.

## Challenges

Tree-based systems have spread rapidly in some dryland zones, but in other zones, adoption continues to lag. Efforts to promote the technology more widely face five major challenges: (1) technical, (2) institutional, (3) legal, (4) economic, and (5) cultural.

**Technical.** The main technical challenge slowing the dissemination of tree-based systems in the drylands is lack of water. Water is needed in all humidity zones during the dry season to maintain tree nurseries, and it is needed throughout the year in the more arid zones to allow watering of recently planted saplings.

**Institutional.** During their early stages of growth, young trees are vulnerable to heavy browsing and to fire. Dissemination of tree-based systems has lagged in areas in which local customs and laws fail to ensure the protection of young trees.

**Legal.** In many dryland countries, forest regulations—even though well intended— discourage farmers from effectively managing indigenous species on their farms. For example, farmers are often required to pay for licenses to cut trees on their own land. Where these policies and regulations have been revised, in most cases farmers have responded with an explosion of tree regeneration on their lands.

**Economic.** The incentives to invest in tree-based systems in the drylands are not often obvious to farmers. Because trees grow slowly in the drylands, the benefits from an investment in trees often take years to materialize. This can be problematic. The long time lag to realize investment returns reduces the attractiveness of tree-based systems, where resource-constrained farmers generally must focus on meeting their families' immediate consumption needs. In the case of trees grown for commercial purposes, a major challenge is the fact that markets for many tree products are as yet poorly developed.

**Cultural.** Despite the accumulating body of evidence demonstrating the benefits of associating trees and crops in dryland zones, and despite the fact that farmers in the drylands have been using tree-based systems for generations, extension messages in many countries continue to encourage farmers to maintain "clean" fields.

## Key messages

Trees can improve the productivity and stability of crop and livestock production systems in the drylands by providing multiple benefits that tend to stand up well in the face of climate shocks.

The importance of tree-based systems and their role will vary depending on the microenvironment.

In **arid zones**, low and uncertain rainfall makes investment in purposefully planted tree-based systems risky. In these zones, FMNR can make sense as a strategy to improve the productivity of pastoral livestock systems. Tree-based systems also have the potential in arid zones to sequester carbon, a function that currently generates little or no revenue for landholders but that could become increasingly important in future with the development of payment for environmental services schemes.

In **semi-arid zones**, tree-based systems have considerable potential to contribute to the productivity, profitability, and sustainability of agro-pastoral systems. Drought remains a threat, however, making some management practices more attractive than others. In semi-arid zones, regenerative tree-based systems should be promoted widely as a foundational practice. In selected areas, particularly areas where irrigation is available, there is scope for purposefully planting trees to produce wood and non-wood products as a way of increasing farmer incomes and improving nutrition.

In **dry subhumid zones**, tree-based systems can perform extremely well, but they are likely to face competition from other agricultural activities and therefore may have difficulty gaining traction. In areas where continuous cereal cropping is taking place, regenerative tree-based systems may be able to help maintain soil fertility. And in areas of higher population density, where markets are well developed, cultivation at small and medium scale of purposefully planted high-value trees could generate significant amounts of income while at the same time contributing to improved nutrition.

# References

Ajayi, O.C., F. Place, F.K. Akinnifesi and G. Sileshi. 2011. "Agricultural Success from Africa: The Case of Fertilizer Tree Systems in Southern Africa (Malawi, Tanzania, Mozambique, Zambia and Zimbabwe)." *International Journal of Agricultural Sustainability* 9(1): 129–36.

Ajayi, O., F. Place, F. Kwesiga, P. Mafongoya. 2007. "Impacts of Improved Tree Fallow in Zambia." In H. Waibel and D. Zilberman, editors, *International Research on Natural Resource Management: Advances in Impact Assessment*. Oxford, UK: Oxford University Press.

Akinnifesi, F., O. Ajayi, G. Sileshi, P. Chirwa, and J. Chianu. 2010. "Fertilizer Tree Systems for Sustainable Food Security in the Maize Based Production Systems of East and Southern Africa Region: A Review." *Agronomy for Sustainable Development* 30: 615–629.

Bayala, J., J. Sanou, Z. Teklahaimanot, A. Kalinganire, and S. Oeudrago. 2014. "Parklands for Buffering Climate Risk and Sustaining Agricultural Production in the Sahel of West Africa." *Current Opinion in Environmental Sustainability* 6: 28–34.

Beedy, T.L., G. Nyamadzawo, E. Luedeling, D-G. Kim, F. Place, and K. Hadgu. 2014. "Carbon Sequestration and Soil Fertility Replenishment by Agroforestry Technologies for Small Landholders of Eastern and Southern Africa." In R. Lal and B.A. Stewart, editors, *Managing Soils of Smallholder Agriculture*. Boca Raton, Florida: CRC Press.

Garrity, D.P., F.K. Akinnifesi, O.C. Ajayi, S.G. Weldesemayat, J.G. Mowo, A. Kalinganire, M. Larwanou, and J. Bayala. 2010. "Evergreen Agriculture: A Robust Approach to Sustainable Food Security in Africa." *Food Security* 2: 197–214.

Glenn, J. V. 2012. "Economic Assessment of Landowner Incentives: Analyses in North Carolina and Malawi." MSc Thesis, Department of Natural Resources, North Carolina State University, Raleigh.

Mafongoya, P.L., E. Kuntashula, and G. Sileshi. 2006. "Managing Soil Fertility and Nutrient Cycles through Fertilizer Trees in Southern Africa." In N. Uphoff, A.S. Ball, E. Fernandes, H. Herren, O. Husson, M. Liang, C. Palm, J. Pretty, P. Sanchez, N. Sanginga, and J. Thies, editors, *Biological Strategies for Sustainable Soil Systems*, 273–89. Taylor & Francis.

Mukandiwa, L., P.H. Mugabe, T.E. Halimani, and H. Hamudikuwanda. 2010. "A Note on the Effect of Supplementing Rangeland Grazing with *Acacia angustissima* Mixed with Pearl Millet on Growth Performance of Goats in a Smallholder Farming Area in Zimbabwe." *Livestock Research for Rural Development* 22: Article #9. http://www.lrrd.org/lrrd22/1/muka22009.htm.

Nair, P.K., V.D. Nair, E.F. Gama-Rodrigues, R. Garcia, S.G. Haile, D. Howlett, B.M. Kumar, M.R. Mosquera-Losada, S.K. Saha, A. Takimoto, and R.G. Tonucci. 2009. "Soil Carbon in Agroforestry Systems: An Unexplored Treasure?" *Nature Precedings*. http://precedings.nature.com/documents/4061/version/1

Orwa, C., A. Mutua, R. Kindt, R. Jamnadass, and S. Anthony. 2009. "Agroforestree Database: A Tree Reference and Selection Guide Version 4.0." World Agroforestry Centre, Kenya.

Place, F., and J. Binam. 2013. "Economic Impacts of Farmer-Managed Natural Regeneration in the Sahel." End-of-Project Technical Report for Free University Amsterdam and International Fund for Agricultural Development (IFAD), World Agroforestry Center, Nairobi, Kenya.

Place, F., D. Garrity, S. Mohan, and P. Agostini. 2016. *Tree-Based Production Systems for Africa's Drylands*. World Bank Studies. Washington, DC: World Bank.

Pratt D.J., F. LeGall, and C. de Haan. 1997. Investing in Pastoralism: Sustainable Natural Resource Use in Arid Africa and the Middle East. World Bank, Washington, DC.

Reij, C., G. Tappan, and M. Smale. 2009. "Agroenvironmental Transformation in the Sahel: Another Kind of Green Revolution." Discussion Paper 00914, International Food Policy Research Institute (IFPRI), Washington, DC.

Sileshi, G., F.K. Akinnifesi, O.C. Ajayi, and F. Place. 2008. "Meta-Analysis of Maize Yield Response to Woody and Herbaceous Legumes in the Sub-Saharan Africa." *Plant Soil* 307: 1–19.

Smith, E. 2010. "Local Knowledge of Natural Regeneration and Tree Management in Sahelian Parklands North of Tominian Mali." MSc thesis, School of Environment, Natural Resources and Geography, University of Bangor, Wales.

## Chapter 7

# Agriculture: More Water and Better Farming for Improved Food Security

*Tom Walker, Christopher Ward, Rafael Torquebiau, Hua Xie,*
*Weston Anderson, Nikos Perez, Claudia Ringler, Liang You,*
*Nicola Cenacchi, Tom Hash, Fred Rattunde, Eva Weltzien,*
*Jawoo Koo, Federica Carfagna, Raffaello Cervigni, Michael Morris*

## Current situation

Agriculture—used here to refer to farming in general and crop cultivation in particular—is one of the two main livelihood strategies practiced in the drylands (the other being livestock-keeping). In the countries of East and West Africa in which drylands are important, agriculture is economically significant, with crop production typically contributing 10–30 percent of GDP and up to 75 percent of agricultural GDP.

Dryland agriculture is diverse, with mixed cropping predominating as a way of protecting against risk. Most dryland farming systems are dominated by one or two main staples, which are grown in association with a range of other crops having dissimilar growth cycles and different maturity dates. Generally speaking, cropping systems in drier areas are dominated by millet and sorghum, due to the superior ability of these two species to tolerate drought and heat. As rainfall levels increase and mean temperatures decline, millet and sorghum give way to maize, which is the dominant crop throughout the wet parts of the semi-arid zone and the subhumid zone. In the wettest part of the drylands, maize is increasingly associated with roots and tubers, including cassava, yam, and sweet potato.

Drylands are generally unfavorable for agriculture. The harsh agro-climatic conditions restrict the potential of many crops, and fields are chronically exposed to unpredictable shocks that can decimate production to the point of causing complete crop loss. The biggest challenge to dryland agriculture is posed by the uncertain availability of water, both in terms of quantity and timing. Although the effects of uncertain and highly variable rainfall can be mitigated through the use of irrigation, irrigation is relatively underdeveloped in the drylands, as it is across the region as a whole. Sub-Saharan Africa has the lowest level of irrigation development in the world. Across the entire region (drylands and non-drylands), about 7.1 million hectares have been developed for irrigation, representing just 3 percent of the total cultivated area. This compares to about 15 percent of cropland that is irrigated worldwide. Not only is irrigation much less developed in Sub-Saharan Africa than elsewhere, but the area that is developed is underused—more than one-fifth of the area equipped with irrigation infrastructure is reported to be out of use. Prospects for catching up with the rest of the world are bleak, as the rate of expansion of new irrigation is slow, averaging about 1 percent per year since 1995.

With irrigation still relatively underdeveloped, crop cultivation in the drylands takes place mainly in rainfed systems. Rainfed crop production in the drylands is highly correlated with rainfall, which is important because drought is a defining feature of the environment (figure 7.1). In most years, farmers sow their crops into dry soil at the beginning of the rainy season, in the expectation that the rains will follow. When the temporal distribution of rainfall differs from expectations, the consequences can be severe. The late arrival of early-season rains may spell crop failure, and terminal drought stress at the end of the growing season can be catastrophic as well. A second major constraint affecting dryland agriculture is extreme temperatures, particularly heat. Although many of the crops grown in drylands have the ability to tolerate wide temperature

**Figure 7.1** Dryland cereal production and rainfall in Burkina Faso, 1960–2000

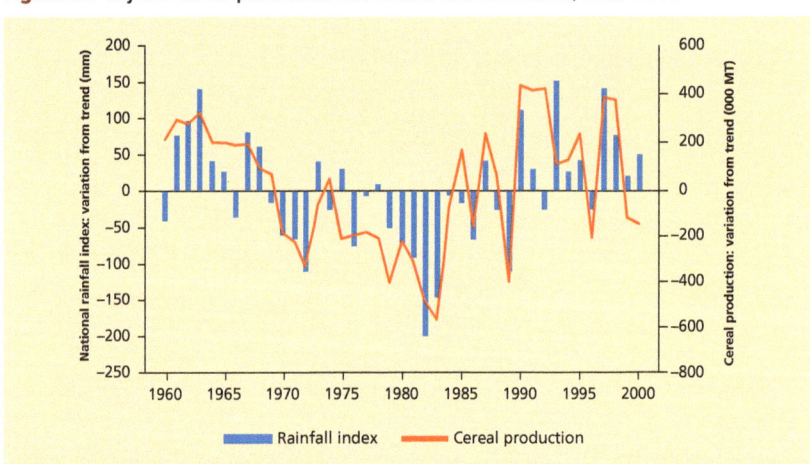

Source: Ward, Torquebiau, and Xie 2016.

Note: mm = millimeters; MT = metric tons.

fluctuations, most are unable to withstand even short periods of extreme heat or cold, especially when these occur at critical stages of the plant growth cycle.

Water scarcity and extreme heat are the two biggest constraints affecting dryland agriculture, but they are hardly the only ones. Low soil fertility and nutrient depletion are chronic problems, with an estimated three-quarters of dryland soils showing symptoms of one of more plant nutrient deficiencies. Eroding winds, uncontrolled burning, and attack by insects such as locusts and army worms can further impair productivity and increase risk in dryland cropping systems.

As a result of the many constraints, productivity in dryland farming systems is generally low, and production tends to fluctuate considerably from year to year. Across all of Sub-Saharan Africa, total factor productivity in agriculture increased very little during the three decades 1960–1990. Not until the mid-1980s did significant numbers of African farmers begin adopting more intensive technologies, leading to a modest acceleration in productivity growth (Fuglie and Rada 2013).

## Opportunities

Despite the challenges they pose to farming, drylands feature a number of agro-climatic conditions that are favorable for plant growth, such as high levels of solar radiation and a relative absence of pests and diseases. These advantages confer possibilities for crop productivity gains. Where there are profitable markets and particularly where farmers have access to reliable water supplies, technological change can occur rapidly, bringing gains in income, reductions in poverty, and increases in resilience.

Agricultural productivity in many parts of the drylands is far below potential, as reflected by large and persistent gaps between yields observed in farmers' fields and yields recorded on experiment stations using optimal levels of inputs and improved management practices. The existence of these yield gaps means that technologies are available with demonstrated capacity to increase and stabilize the productivity of dryland agriculture. These technologies are not all the same, however; the benefits they deliver depend on the degree to which they address each of the three determinants of vulnerability and resilience (exposure, sensitivity, coping capacity).

### Reducing exposure

Unlike the case of many livestock-keepers who can move their herds to avoid exposure to droughts and other related shocks, farmers cannot move their fields. For this reason, agriculture will always be exposed to weather shocks, especially droughts.

## Reducing sensitivity

Farmers living in dryland regions are affected by droughts only to the extent that their farming activities are sensitive to the effects of those droughts. For this reason, interventions that reduce the sensitivity of dryland agriculture to droughts have the potential to reduce the vulnerability and improve the resilience of households that depend on farming as their principal livelihood source. Two broad categories of interventions are distinguished here that reduce sensitivity of crop farming to droughts: (1) improved management practices for rainfed agriculture, and (2) irrigation development.

## Improved management practices for rainfed agriculture

Where there are profitable markets and particularly where farmers have access to reliable water supplies, technological change in rainfed cropping systems can occur rapidly, bringing gains in income, reductions in poverty, and increases in resilience. The modeling exercise done for this book confirmed that several opportunities for accelerating the pace of technological change offer particularly bright prospects, described as follows.

**Accelerating the rate of varietal turnover.** Modern varieties (MVs) of cereals, such as rice, wheat, and maize, played a major role in driving the Green Revolutions of Asia and Latin America, but they have had much less impact in Sub-Saharan Africa, where adoption of MVs has lagged (Walker et al. 2014). In 2010, across Sub-Saharan Africa as a whole, the average rate of MV adoption among 20 field crops stood at around 35 percent (figure 7.2). While this

**Figure 7.2** Adoption rate of modern varieties by crop in Sub-Saharan Africa, 2010 (%)

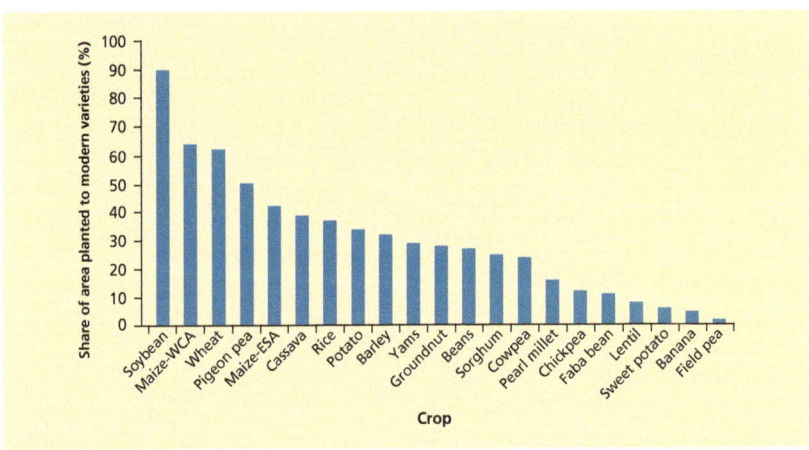

*Source:* Constructed from Walker et al. 2014.
*Note:* Maize is divided into East and Southern Africa (ESA) and West and Central Africa (WCA).

adoption rate is considerably lower than the rate achieved in other developing regions, the uptake of MVs in Africa has accelerated in recent years, particularly in the case of maize and cassava, the leading dryland cereal and root crops (Walker and Alwang 2015). If current adoption rates continue, two-thirds of dryland areas will be sown to MVs by 2030.

**Increasing the availability of hybrids.** Thanks to the phenomenon of heterosis (commonly known as "hybrid vigor"), well-adapted hybrids have two main advantages over well-adapted improved varieties: higher yield potential and greater yield stability. In addition, because these advantages of hybrids are assured only when farmers purchase new seed for every cropping cycle, the demand for hybrid seed tends to be strong, creating incentives for private companies to make sure that the market is well supplied. Yet despite the superior performance of hybrids and the stronger incentives for seed companies, adoption of hybrids remains low in many dryland regions, and hybrid seed remains scarce in local markets. Increasing the availability of hybrids could increase resilience, especially for maize, sorghum, and pearl millet in West Africa, which account for about 40 percent of dryland cropped area in Sub-Saharan Africa.

**Improving fertility management.** Because low soil fertility constitutes a major constraint to farming in the drylands, diffusion of improved fertility management practices is essential for the sustainable intensification of dryland agriculture. A number of practices have demonstrated their effectiveness under diverse dryland conditions, including mulching, green manuring, composting, intercropping with legumes, and judicious use of mineral fertilizer (for a summary, see Walker and Alwang 2015). The impact of improved soil fertility management technologies is amplified when MVs are introduced at the same time because of the synergistic effects between improved germplasm and improved management practices.

**Improving agricultural water management.** In the drylands, which are characterized by conditions of chronic water scarcity and climatic unpredictability, soil moisture is often inadequate to achieve a decent yield, and in times of drought, farmers may face total crop failure. Many households in the drylands that rely on farming as their primary livelihood strategy are highly sensitive to soil moisture risk and to the resulting low yields or crop failure. Therefore increasing the availability of water and improving the efficiency with which available water is used can have a transformational impact in dryland rainfed agriculture.

Agricultural water management in dryland environments aims to reduce sensitivity to drought and strengthen coping capacity by bringing moisture to the plant root zone in the right quantity and quality and at the right time to achieve higher levels of productivity, essentially by one or more of three routes: (1) "just-in-time" watering to bridge drought gaps and save the crop; (2) delivering quality water at optimal intervals to the plant root zone through good water service to the field and in-field water management to promote optimal plant growth; and (3) combining water management with soil and crop management to achieve optimal crop water productivity (Ward, Torquebiau, and Xie 2016).

The most secure way to increase the availability of water to growing plants is through irrigation development, discussed in the next section (see box 7.2). Short of irrigation development, however, many tried and tested technologies are available to improve water availability and management for rainfed farming in the drylands (table 7.1). Investment programs in areas where full irrigation is not an option should focus on improved agricultural water management as part of a total livelihood package.

**Technologies to improve crop productivity in dryland environments.** A technology assessment carried out for this book identified five technologies with demonstrated capacity to improve crop productivity in dryland environments: (1) drought-tolerant improved varieties, (2) heat-tolerant improved varieties, (3) fertilizer, (4) water harvesting, and (5) farmer-managed natural regeneration (FMNR) of indigenous trees. The ability of these five technologies to reduce vulnerability and increase resilience of agriculture-dependent

**Table 7.1** Water management strategies for rainfed agriculture

| Aim | Strategy | Purpose | Techniques and structural measures |
|---|---|---|---|
| Improve water use efficiency by increasing water available to the plant roots | Soil and water conservation | Concentrate rainfall around crop roots | Bunds, ridges, broad-beds and furrows, micro basins, runoff strips<br><br>Planting pits |
| | | Maximize rainwater infiltration | Terracing, contour cultivation, conservation agriculture, dead furrows, staggered trenches |
| | Evaporation management | Reduce non-productive evaporation | Dry planting, mulching, conservation agriculture, inter-cropping, windbreaks, agroforestry, early plant vigor, vegetative bunds |
| | Water harvesting | Mitigate dry spells with supplementary irrigation, protect springs, recharge groundwater, enable off-season irrigation, and permit multiple uses of water | Surface micro dams, subsurface tanks, farm ponds, percolation dams and tanks, diversion and recharging structures |
| Improve water productivity by increasing productivity per unit of water consumed | Integrated soil, crop, and water management | Increase proportion of evapotranspiration flowing as productive transpiration and so obtain "more crop per drop" | Increase plant water uptake capacity through conservation agriculture, dry planting (early), improved crop varieties, optimum crop spacing, soil fertility management, optimum crop rotation, intercropping, pest control, and organic matter management |

households was assessed through a multi-step modeling exercise using the "Africa RiskView" (ARV) model developed by the African Risk Capacity (a specialized agency of the African Union) and the Decision Support System for Agrotechnology Transfer (DSSAT) crop simulation model. The ARV model uses drought vulnerability profiles of the population against which drought impacts are calculated to estimate drought-affected populations under different drought scenarios. The DSSAT model allows estimation of the impacts on yields of indicator crops—in this case millet, sorghum, and maize—of the five technologies under different drought scenarios.

By combining the results of the two models and including screening criteria to restrict adoption of each technology to zones in which adoption would likely be profitable for the farmer, it was possible to estimate the number of households living in drylands in 2030 that would be made resilient by adopting one of the technologies. (A more detailed description of the modeling approach appears in the Appendix.) Because simultaneous adoption of two or more technologies results in interactive effects that are difficult to capture in the DSSAT model, the impacts of the best-bet technologies were modeled separately, and only the most effective technology was assumed to be adopted in each location. Thus the results of the modeling exercise are conservative, because they do not allow for simultaneous adoption of multiple technologies, which is likely to occur in many situations.

The results of the simulation exercise are summarized in figure 7.3. Overall, improving soil fertility through application of fertilizer was found to have the greatest potential for increasing resilience in the drylands. After soil fertility management, the technologies with the next greatest potential for increasing resilience were found to be drought-tolerant germplasm and FMNR of indigenous tree species. The effectiveness of the latter technology increases with tree

**Figure 7.3** Contribution of improved cropping technologies to reducing vulnerability (%)

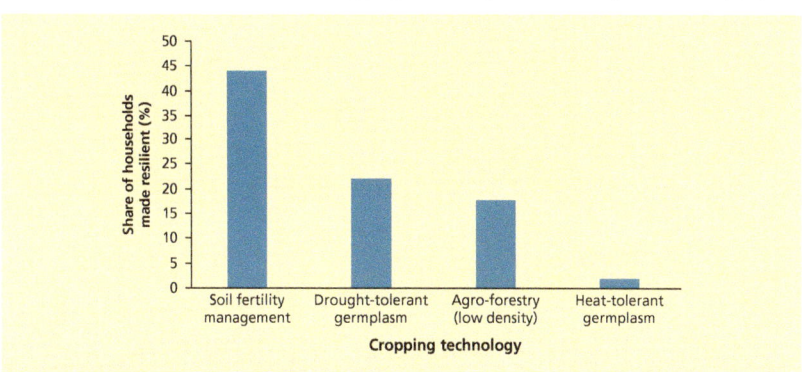

*Source:* Calculation based on the approach discussed in the Appendix.

density: establishing and maintaining 10 trees per hectare (ha) on average was found to be significantly more beneficial than establishing and maintaining 5 trees per hectare on average. Heat-tolerant germplasm in and of itself was found to have limited potential. Water harvesting practices were found to have limited potential, due to the relatively high cost compared to the limited expected returns from higher yields.

In summary, adoption of improved cropping technologies could make an important contribution to reducing vulnerability and increasing resilience in the drylands, particularly in countries in which a large proportion of vulnerable households depend on agriculture as a major livelihood source. Figure 7.4 shows the reduction in 2030 in the share of drought-affected households relative to the business as usual (BAU) scenario that would occur if the most effective technology were adopted in every location in which adoption would be profitable. Across the drylands as a whole, just under 20 percent of all households could be made resilient by adopting one or more of the improved cropping technologies. In some countries the share would be much higher. For example, in Ethiopia one-half of the drought-affected households could be made resilient in the face of drought by adopting improved cropping technology. In Senegal and Niger more than one-quarter of the drought-affected households could be made resilient.

## Irrigation development

The most reliable way to reduce the sensitivity of cropping systems in the drylands to drought shocks and to ensure adequate water supplies at critical periods during the cropping season is through irrigation. Despite their prevailing aridity, many dryland areas have considerable water resources that can be used

**Figure 7.4** Reduction in the share of drought-affected households from adoption of improved cropping technologies relative to BAU scenario, 2030 (%)

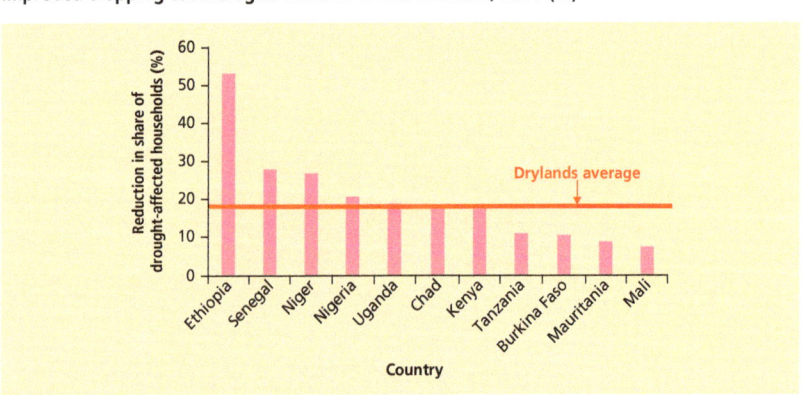

Source: Calculation based on the approach discussed in the Appendix.

for irrigation, both surface water and groundwater. Yet much of this potential remains unexploited: dryland countries have developed less than one-third of their technical irrigation potential, and more than one-fifth of the area developed for irrigation is currently not in use (Xie et al. 2015).

## Small-scale irrigation

Because of its relative affordability and manageability, small-scale irrigation arguably offers the most important opportunities to improve agricultural water management in drylands. Modeling work carried out for this book suggests that using conservative assumptions about costs and returns to investment capital allows considerable scope for further development of small-scale irrigation in dryland regions of Africa—up to 3 million hectares or even more (Xie et al. 2015).

Individual smallholder irrigation using low-cost pumps is spreading fast in many dryland regions, drawing water from both groundwater and surface sources. Because of the recurrent cash outlays needed to pay for fuel, operation and maintenance, and production inputs such as seed and fertilizer, small-scale irrigation works best when cash crops are being produced and when farmers have ready access to nearby markets where they can sell their production (Ward, Torquebiau, and Xie 2016).

In addition to individual smallholder irrigation, community-based small-scale irrigation offers considerable scope for expansion in drylands. Small-scale community-based irrigation has expanded in recent decades in response to new market opportunities, often with support from development programs. Because it is essentially farmer-managed, community-based small-scale irrigation tends to be well adapted to local biophysical conditions and socioeconomic circumstances (Ward, Torquebiau, and Xie 2016).

## Large-scale irrigation

Large-scale irrigation offers additional opportunities for increasing and stabilizing agricultural production in dryland areas. It is difficult to predict to what extent these opportunities will be exploited, however. Because the benefits generated by agriculture alone rarely justify the cost of constructing large dams, future growth in large-scale irrigation will likely depend on decisions to invest in dams whose primary function is to generate hydro-power.

Some opportunities are more accessible than others. For example, there is scope to double production in the drylands of existing large-scale irrigation schemes that currently are underutilized. Technical and institutional modernization on the 5 million hectares currently being irrigated in the drylands could greatly increase yields—even double them in some cases—at an average cost about US$2,700 per hectare, less than half the cost of developing new irrigation. In addition, there is scope for bringing back into production some of the more than one million hectares in the drylands that are equipped for irrigation but not currently being irrigated.

With respect to developing new large-scale irrigation in the drylands, much will depend on future investments in the energy sector. It is beyond question that the technical potential to expand large-scale irrigation is significant, if technical potential is defined in terms of the availability of water and arable land. Taking into account the 120 large dams currently in existence or included in national development plans in dryland countries, and assuming conservatively that 30 percent of the water stored in these dams will be available for irrigation, up to 1.5 million hectares could be developed for large-scale irrigation. Approximately two-thirds of these dams are already operational, meaning only conveyance and distribution systems and pumping equipment are needed to bring water to the fields.

## Irrigation development potential in drylands through 2030

What might be the potential impact by 2030 on productivity and production if the potential for small-scale and large-scale irrigation in the drylands was fully developed? This question was explored using a modeling approach described in Xie et al. (2015). As a baseline, it was estimated that in 2000 approximately 6.43 million hectares in all of Sub-Saharan Africa were equipped for irrigation. Of this area, approximately 4.56 million hectares (71 percent) were located in dryland regions. Table 7.2 shows the additional area that could potentially be developed for large-scale and small-scale irrigation by 2030, assuming moderate capital investment costs and two minimum acceptable internal rates of return (IRR). Restricting the analysis to the drylands as they are defined in this book (Aridity Zones 3–6), depending on the assumptions, by 2030 as little as 3.9

**Table 7.2** Irrigation development potential by 2030, by aridity zone (hectares)

| | Total cropland | Large-scale irrigation | | Small-scale irrigation | |
|---|---|---|---|---|---|
| | | 5% IRR | 12% IRR | 5% IRR | 12% IRR |
| Hyper-arid (Zone 1) | 1,248,862 | 60,170 | 47,624 | 0 | 389 |
| Arid (Zone 2) | 567,069 | 0 | 0 | 2,910 | 5,732 |
| **Dry Semi-arid (Zone 3)** | 16,308,307 | 91,926 | 96,428 | 307,768 | 142,116 |
| **Wet Semi-arid (Zone 4)** | 25,127,335 | 141,132 | 95,102 | 1,238,674 | 897,492 |
| **Dry Subhumid (Zone 5)** | 29,546,353 | 240,395 | 182,831 | 1,716,223 | 1,242,597 |
| **Wet Subhumid (Zone 6)** | 35,610,403 | 450,073 | 373,914 | 1,891,591 | 1,620,670 |
| Humid (Zone 7) | 76,139,002 | 713,412 | 499,563 | 3,121,388 | 2,931,384 |
| Grand Total | 184,547,331 | 1,697,108 | 1,295,462 | 8,278,554 | 6,840,381 |

*Source:* You, Wood, and Wood-Sichra 2009; Xie et al. 2015.

*Note:* IRR = internal rate of return.

million hectares or as much as 5.2 million hectares could be developed for small-scale irrigation. For large-scale irrigation development, which will depend on the rehabilitation and construction of dams, the area is much more limited, ranging from a low of 0.75 million hectares to a high of 0.92 million hectares.

In dryland regions of West Africa and East Africa, prospects for irrigation development by 2030 vary considerably by country (table 7.3). It is interesting to note, however, that even if the potential for irrigation were fully exploited, in most cases the irrigated area would comprise only 3–20 percent of total cropland.

The potential for irrigation development in dryland regions of Sub-Saharan Africa is summarized in map 7.1. Two clear messages emerge from the

**Table 7.3** Irrigation development potential by 2030, East and West Africa (moderate cost and 5% IRR)

|  | Cropland in 2000 (ha) | Irrigated area potential (ha) | Irrigated area potential (% of cropland) |
|---|---|---|---|
| Nigeria | 24,523,253 | 1,617,654 | 7 |
| Ghana | 1,759,898 | 312,275 | 18 |
| Senegal | 2,266,221 | 255,901 | 11 |
| Burkina Faso | 5,176,476 | 174,513 | 3 |
| Mali | 4,696,988 | 141,362 | 3 |
| Chad | 3,539,511 | 94,080 | 3 |
| Niger | 12,232,511 | 118,795 | 1 |
| Benin | 2,030,091 | 135,989 | 7 |
| Mozambique | 2,601,577 | 76,433 | 3 |
| Côte d'Ivoire | 968,534 | 74,316 | 8 |
| Mauritania | 284,483 | 100,340 | 35 |
| Togo | 790,188 | 61,798 | 8 |
| Cameroon | 1,145,331 | 56,664 | 5 |
| Guinea | 214,349 | 22,927 | 11 |
| Gambia, The | 277,146 | 17,682 | 6 |
| Kenya | 2,629,859 | 335,705 | 13 |
| Ethiopia | 4,801,840 | 245,629 | 5 |
| Somalia | 935,603 | 230,028 | 25 |
| Eritrea | 669,799 | 27,865 | 4 |
| Swaziland | 95,822 | 13,488 | 14 |
| Sudan | 10,449,867 | 11,775 | 0.1 |
| Djibouti | 5,051 | 3,648 | 72 |

*Source:* Xie et al. 2015.

modeling work. First, the potential for irrigation development in the drylands is substantial, but the likely impact on crop production pales in comparison to the impact on crop production that could be achieved by fully exploiting the available opportunities to develop rainfed agriculture. Rainfed agriculture is far more important than irrigated agriculture in the drylands and will remain that way for the foreseeable future. Second, within the irrigation sector, although large-scale irrigation is generally the most reliable form of irrigation, compared to small-scale irrigation, large-scale irrigation has a limited area potential and is much smaller in size. The reason is that expansion of large-scale irrigation will depend on investments in dams that will be rehabilitated or constructed for purposes other than agriculture.

To summarize, the modeling work done for this book suggests that there is considerable scope for increasing the irrigated area in the drylands. Development of a further 6.1 million hectares (in addition to the current 4.6 million hectares) is technically feasible and economically justifiable (Xie et al. 2015). Overall, irrigation development could have a large, possibly transformational impact on farming systems and on resilience. Prospects are brightest for small-scale irrigation because of its lower costs, more decentralized management, and likely higher levels of farmer participation.

There is also considerable potential for large-scale irrigation development, concentrated along corridors located downstream from dams that will have been constructed for other (non-agricultural) purposes. Investment costs for

**Map 7.1** Potential for development of small- and large-scale irrigation in Sub-Saharan Africa

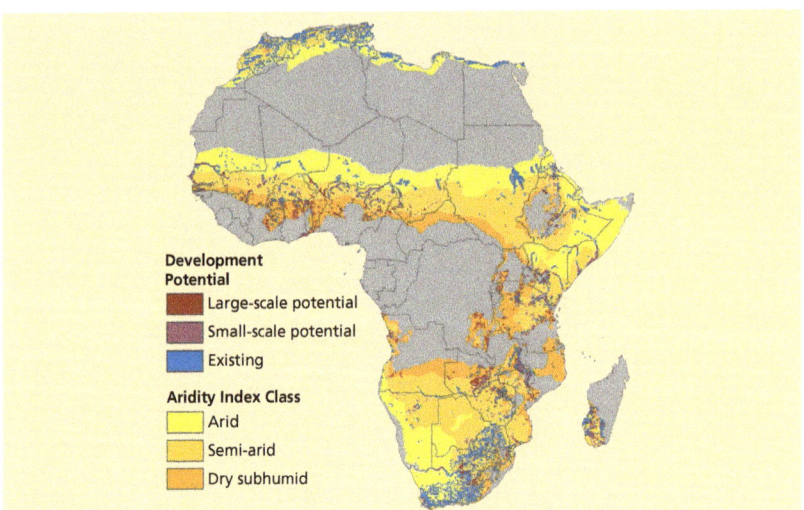

*Source:* © IFPRI. Reproduced with permission from Xie et al. 2015; further permission required for reuse.

large-scale irrigation are roughly three times higher than for small-scale irrigation, but the value of the incremental production and the amount of employment created are three times as great. Large-scale irrigation poses technical, economic, and institutional challenges and risks, however, so investment in large-scale irrigation schemes is likely to proceed slowly, to provide time for models to be worked out that ensure that such schemes can be operated profitably and sustainably. In some dryland countries, improving existing large-scale irrigation schemes may pay higher returns than building new schemes.

Small- and large-scale irrigation schemes both have the potential to contribute to increased resilience, but their contributions will be somewhat different (Ward, Torquebiau, and Xie 2016). Small-scale irrigation in the drylands is used in a wide range of mixed farming systems, helping to raise and stabilize crop yields and thereby allowing large numbers of poor households to grow more home-consumed food and increase their income from cash sales. In contrast, large-scale irrigation is often associated with specialized production systems that feed into distinct and separate value chains, so it strengthens resilience by allowing households to generate cash incomes that are relatively insensitive to shocks. While the number of households that can be accommodated on large-scale schemes is usually limited, such schemes tend also to create new employment opportunities for wage laborers, thereby enhancing resilience for a broader segment of the rural population.

### Improving coping capacity

Agricultural households that live in dryland regions, being unable to move out of harm's way when shocks occur and having livelihoods that are sensitive to shocks, suffer frequent income losses. For these households, the ability to survive will depend mainly on their coping capacity, that is, on their ability to draw on their own accumulated resources or resources provided by others to meet their needs during a critical period until their livelihood strategies can be reestablished.

**Public policy interventions.** Experience suggests that many agricultural households when hit by a shock soon exhaust their limited accumulated resources, leaving them critically dependent on public programs. Public policy thus plays an important role in supporting the recovery process, particularly for non-resilient households. In considering the instruments available to the government, it is useful to distinguish between interventions that can be implemented relatively quickly versus interventions that require time to produce results.

Public policy interventions that can be implemented in the short run to strengthen the coping capacity of agriculture-dependent populations include: (1) introducing crop insurance to provide compensation for production losses, and (2) establishing scalable safety nets to provide alternative sources of income until the farming enterprise can be fully restored. (Crop insurance is discussed in the next section; scalable safety nets are discussed in Chapter 9.)

**Crop insurance.** In theory, crop insurance addresses the problem of systemic risk from yield variability in dryland agriculture (for a general discussion, see Hazell, Pomareda, and Valdes 1986). In addition to directly protecting farmers from yield losses due to adverse weather and outbreaks of diseases or pests, crop insurance indirectly enhances resiliency in the production environment because farmers with insurance will be more willing to adopt technologies perceived to be profitable without having to worry as much about the vagaries of the weather. Increased profits attributable to the improved technologies can be reinvested to further limit sensitivity to risk and to improve coping mechanisms and strategies.

In practice, crop insurance that is voluntary and oriented to the individual producer is vulnerable to consistent and sizable losses because of moral hazard and adverse selection (Brown, Mobarak, and Zolanska 2014). Moral hazard refers to negative incentives as farmers are rewarded for exerting less effort when yields approach payout trigger points. Adverse selection becomes a problem when the more productive farmers with higher yields do not participate in crop insurance programs. Both moral hazard and adverse selection erode the actuarial basis for cost-effective crop insurance.

Interest in crop insurance has waxed and waned over the years, and a number of pilot programs have been launched to test out different design features. Several advances in design have brightened the prospects for crop insurance, including the use of a homogeneous area approach to compensation (which eliminated the moral hazard problem, because individual farmers could not manipulate yield estimates calculated over large areas) and, instead of targeting individual farmers, using rainfall instead of yields as the criterion for determining payouts (much easier to measure, because rainfall data are more readily available than yield data, especially following the advent of automatic weather stations).

Despite some isolated success stories, demand for crop insurance among smallholder farmers has remained weak, even when rainfall insurance has been partially subsidized and thoroughly explained. With sufficient targeting and structuring of design to highly contextual conditions, the widely acknowledged problem of weak demand may not be insurmountable. But at least one seasoned observer of insurance over the past 40 years believes that rainfall insurance is not a viable option for improved risk management drylands, arguing that poor farmers are often cash/credit-constrained and therefore cannot advance the money before sowing time to buy insurance that pays out only after the harvest (Binswanger-Mkhize 2012). Others are more sanguine. For example, Brown, Mobarak, and Zelanska (2014) concede that credit constraints, limited financial literacy of farmers, and basis risk limit the demand for rainfall insurance, but they are confident that as the chain of evidence becomes longer with research and as several of these obstacles are overcome, pilot applications can be scaled up to make a substantive contribution to risk management in dryland agriculture.

## Challenges

Opportunities exist to increase and stabilize agricultural production in drylands, but they will not be easy to exploit. Multiple constraints will have to be overcome to enable the successful adoption of productivity-enhancing improved technologies.

The first and perhaps most obvious challenge is financial. Adoption of improved agricultural technology entails two types of cost: (1) costs incurred by the farmers who adopt the technology, especially the costs of purchased inputs such as improved seed, fertilizer, and crop chemicals; and (2) costs incurred by the public sector in promoting the improved technology (e.g., the cost of paying extension agents to provide advisory services, mount publicity campaigns, and train farmers in the use of new technologies). Depending on the technology, the first type of cost can be small (e.g., in the case of improved seed) or large (e.g., in the case of fertilizer). In the latter case, farmers may lack the resources to pay, in which case adoption is unlikely to occur without subsidies or other forms of assistance. The second type of cost is generally quite modest compared to other types of public interventions, as large numbers of farmers can often be reached through relatively low-cost promotional campaigns. Estimates made for this book suggest that five cropping technologies—drought-tolerant varieties, heat-tolerant varieties, chemical fertilizer, water harvesting, and tree-based systems—could be promoted throughout the dryland countries in East and West Africa for US$126–426 million, depending on how effectively promotional efforts are targeted (for details, see Walker et al. 2016). It is important to note, however, that just because an improved technology has been promoted does not mean it will be adopted, as farmers reached by the extension campaign will have to weigh numerous factors before deciding whether or not a promoted technology is right for them.

Aside from cost, several other types of challenges will have to be overcome to ensure successful uptake of improved agricultural production technologies.

**Harsh agro-climatic conditions.** Improved crop production technologies can deliver significant benefits during years of normal weather, but even the best technologies are likely to fail in the face of prolonged drought or extreme heat (see box 7.1). In dryland areas where extreme weather events are common, investing in improved technologies carries risk, which some farmers—particularly the poorest farmers—may be unwilling to take on.

**Infrastructure constraints.** Farmers will be willing to invest in improved technologies only when they are confident that they will be able to produce a decent crop and sell surplus production for remunerative prices. In the drylands

## How will climate change affect dryland agriculture?

The effects on dryland agriculture of droughts and other extreme weather events are readily apparent. In contrast, the effects of climate change resulting from global warming are much less visible, since they occur gradually and manifest themselves differentially through space and time.

Lobell and Field (2007) carried out a comprehensive review of crop modeling exercises and climatic analyses on the impacts of global warming on crop productivity. Maize was identified as the crop most requiring attention in Sub-Saharan Africa, due to its economic and nutritional importance. Because maize germplasm is sensitive to temperature changes, Lobell and Field concluded that the relevant question is not whether climate change will have deleterious impacts on maize yields, but rather how much productivity will be lost from rising temperatures.

Fischer, Byerlee, and Edmeades (2014) recently addressed this issue in a survey of the burgeoning literature on the agricultural consequences of global warming. Among their conclusions:

- $CO_2$ is expected to increase by 26 percent to 480 parts per million (ppm) by 2050; with rising $CO_2$ levels, average global temperatures are forecast to increase by 2°C by 2050.

- Chronic warming, especially hot spells above 30°C, depresses yields by speeding up crop development and by reducing grain numbers and size.

- Predicted changes in precipitation attributed to global warming are not that sizable and are too uncertain to warrant rigorous impact assessment at this time.

- Estimates from regression studies and simulation modeling suggest that average yields of maize, rice, and wheat will fall by 5 percent for each 1°C increase in temperature. In the absence of adaptation, global warming of 2°C by 2050 would lead to a 10 percent decrease in cereal yields.

- These pure temperature effects will be offset by gains from increasing $CO_2$ concentration especially in crops that use the C3 carbon fixation metabolic pathway (wheat, rice, and soybean), where the utilization of $CO_2$ in photosynthesis is not as efficient as in coarse cereals. Hence, the total yield effect is negligible in rice and wheat and is equivalent to an 8 percent loss of productivity in maize.

Fischer, Byerlee, and Edmeades (2014) express optimism that plant breeding and crop agronomy can be deployed to dampen declines in yield from global warming by 2050. Many agricultural research centers have committed

(continued next page)

**Box 7.1** *(continued)*

resources to screening materials for and finding sources of tolerance to heat stress. In the past, sustained breeding efforts for resistance or tolerance to heat stress have not paid dividends for maize at CIMMYT (International Maize and Wheat Improvement Center) and for potato at CIP (International Potato Center). However, several physiological aspects remain to be explored, which could provide the basis for effective heat tolerance in these and other crops. Pearl millet, a hybrid prized for its heat tolerance, is one of the leading cultivars in India in the State of Rajasthan, an environment very similar to the Southern Sahelian Zone (Asare-Marfo et al. 2013).

Fischer, Byerlee and Edmeades (2014) also highlight tactical crop management as a source of innovations designed to combat the adverse effects of climate change. The opportunities are largely location-specific and rely on knowledge about timeliness in the use of inputs. Many of the advances in this area will likely be made in the course of "normal" crop improvement, as the additionality of global warming to the other problems being addressed by agronomic researchers is difficult to envisage in highly specific terms.

producing a decent crop and selling surplus production at remunerative prices are often threatened by underdeveloped irrigation infrastructure, inadequate and unreliable power supplies, and weak transport systems.

**Institutional weaknesses.** The development of improved agricultural technologies and the transfer of these technologies to farmers are joint-impact, high-exclusion-cost activities, which is why they are usually considered public goods and provided through public institutions. Yet in most dryland countries, the public institutions that provide research and extension services are weak and ineffective. Provision of production inputs (e.g., seed, fertilizer, crop chemicals) and financial services are activities that lend themselves more readily to private provision, but the riskiness and low profitability of dryland agriculture has discouraged investment by private firms, so distribution networks for inputs remain underdeveloped, and financial institutions lending to the agriculture sector are few and far between.

**Economic constraints and trade-offs.** The low productivity of dryland agriculture is compounded by the lack of economic incentives to invest in the sector. With production dispersed across vast areas, value chains poorly articulated and inefficient, and agricultural policies fragmented and often acting at cross purposes, dryland agriculture faces a number of daunting economic constraints and trade-offs (box 7.2).

**BOX 7.2**

## Rainfed or irrigated agriculture: A fundamental choice

In seeking to improve the productivity, stability, and sustainability of dryland agriculture, policy makers face a fundamental question: Should attention be focused on improving rainfed production systems, expanding irrigated production systems, or both?

Currently, more than 90 percent of the staples produced and consumed in Sub-Saharan Africa is produced in rainfed systems, and only 5 percent is produced under irrigation. Using realistic assumptions about future area expansion and yield growth, the UN Food and Agriculture Organization (FAO) projects that rainfed agriculture can continue to meet 90 percent of incremental demand for decades to come. Noting that investment in irrigation is economically justifiable only when irrigation facilities can be used to produce high-value cash crops, FAO projects that as soon as 2050, irrigated production is unlikely to contribute more than 10 percent of staples production.

The FAO vision, which is shared by many analysts, suggests that African policy makers and development partners should follow a strategy of promoting production of cereals and grain legumes in drylands, and rice and horticultural crops in irrigated zones. Investments should be tailored accordingly. In zones deemed unfavorable for irrigation, efforts should focus on promoting adoption of improved technologies that can improve productivity and stabilize production of rainfed agriculture, with an emphasis on reducing risk and increasing resilience among vulnerable households. In zones deemed favorable for irrigation, efforts should focus on developing irrigation and promoting production of high-value crops, with emphases on increasing revenues, improving food security, and reducing poverty.

Deciding an appropriate balance between these two complementary objectives will not be easy. From a public policy perspective, given a fixed amount of resources, there is a clear trade-off between investing in small improvements for the large number of households in the drylands that engage in rainfed production, and investing in large improvements for the relatively small number of households that could take advantage of irrigation technology. Investments that target rainfed production systems will not promote highly visible results, but because they can benefit so many households, they have the potential to improve the livelihoods and increase the resilience of the large majority of the population. The policy choice thus pits small reductions in poverty for the many against large reductions in poverty for the few. And given the vast discrepancy in the numbers of households falling into each category, as well as the high cost of irrigation development, targeting dryland agriculture is likely to be the better choice.

## Key messages

More than 200 million people living in dryland regions of Sub-Saharan Africa make their living from agriculture. Most of these people are exposed to weather shocks, especially drought, that can decimate their incomes, destroy their assets, and plunge them into a poverty trap from which it is difficult to emerge. Their lack of resilience in the face of these shocks can be attributed in large part to the poor performance of agriculture on which their livelihood depends.

Opportunities exist to improve the fortunes of these households. Improved farming technologies are available that can increase and stabilize the production of millet, sorghum, maize, and other leading staples. Yet most of these technologies have not been adopted on a large scale, for reasons that include lack of farmer knowledge, non-availability of inputs, unfavorable price incentives, and high levels of production risk.

Irrigation is technically and economically feasible in some areas and offers additional opportunities to increase and stabilize food production, especially in the case of small-scale irrigation systems, which tend to be more affordable and easier to manage. Large-scale, dam-based irrigation systems make sense in certain situations, but their potential is more difficult to exploit because of high investment costs and daunting institutional and governance challenges. While irrigation represents an excellent option in some areas, it is important to keep in mind that prospects for irrigation development are limited in the drylands, so for the foreseeable future, rainfed agriculture will continue to be far more important.

Future production growth in the drylands is expected to come mainly from raising yields and increasing the number of crop rotations on land that is already being cultivated (intensification), rather than from bringing new land into cultivation (extensification). Controlling for rainfall, average yields in rainfed cropping systems in Sub-Saharan Africa are still much lower than yields in rainfed cropping systems in other regions, suggesting that there is considerable scope to intensify production in these systems. Furthermore, unlike in other regions, production of low-value cereals under irrigation is not generally economic in Sub-Saharan Africa unless the cereals can be grown in rotation with one or more high-value cash crops. The long-run strategy for dryland agriculture therefore must be to promote production of staples in rainfed systems and production of high-value cereals (e.g., rice), horticultural cops, and industrial crops in irrigated systems.

Considerable potential exists in the drylands to improve the productivity of rainfed agriculture and to expand irrigation. Exploiting the available opportunities will require policy reforms and institutional changes backed by supporting investment. Attention must focus on:

- Strengthening innovation systems at the national and regional level, for example, by supporting the emergence of multi-actor networks that can leverage the strengths of public institutions, private firms, and civil society organizations.
- Promoting improvements in rainfed agriculture to increase and stabilize production of food staples and strengthen resilience of vulnerable households.
- Promoting investments in irrigated agriculture, both small-scale and large scale, to increase production of high-value cash crops and raise incomes and reduce poverty of commercially oriented farmers.

Improving the productivity and stability of agriculture in the drylands has the potential to make a significant contribution to reducing vulnerability and increasing resilience. At the same time, it is important to keep in mind that in an environment characterized by limited agro-climatic potential and subject to repeated shocks, farming on small land holdings may not generate sufficient income to bring people out of poverty.

## References

Asare-Marfo, D., E. Birol, B. Karandikar, D. Roy, and S. Singh. 2013. "Varietal Adoption of Pearl Millet (Bajra) in Maharashtra and Rajasthan, India: A Summary Paper." HarvestPlus, International Food Policy Research Institute, Washington, DC.

Binswanger-Mkhize, H.P. 2012. "Is There Too Much Hype about Index-Based Agricultural Insurance?" *Journal of Development Studies* 48: 187–200.

Brown, J.K., M.A. Mobarak, and T.V. Zelanska. 2014. "Barriers to Adoption of Products and Technologies That Aid Risk Management in Developing Counties." Background paper to the *World Bank 2014 Development Report*. World Bank, Washington, DC.

Fischer, R.A., D. Byerlee, and G.O. Edmeades. 2014. "Crop Yields and Global Food Security: Will Yield Increase Continue to Feed the World?" ACIAR Monograph No. 158. Australian Centre for International Agricultural Research (ACIAR): Canberra. http://aciar.gov.au/publication/mn158.

Fuglie, K.O., and N.E. Rada. 2013. "Resources, Policies, and Agricultural Productivity in Sub-Saharan Africa." ERR-145, U.S. Department of Agriculture, Economic Research Service, Washington, DC.

Hazell, P., C. Pomareda, and A. Valdes, eds. 1986. *Crop Insurance for Agricultural Development: Issues and Experience*. Baltimore, MD: Johns Hopkins University Press.

Lobell, D.B., and C.B. Field. 2007. "Global Scale Climate-Crop Yield Relationships and the Impacts of Recent Warming." *Environmental Research Letters* 2 (1).

Walker, T., with T. Hash, F. Rattunde, and E. Weltzien. 2016. *Improved Crop Productivity for Africa's Drylands*. World Bank Studies. Washington, DC: World Bank.

Walker, T., and J. Alwang, eds. 2015. *Improved Varieties in the Food Crops of Sub-Saharan Africa: Assessing Progress in the Generation, Adoption and Impact of New Technologies*. Wallingford, UK: CABI Publishing.

Walker, T., A. Alene, J. Ndjeunga, R. Labarta, Y. Yigezu, A. Diagne, R. Andrade, R. Muthoni, H. De Groote, K. Mausch, C. Yirga, F. Simtowe, E. Katungi, W. Jogo, M. Jaleta, and S. Pandey. 2014. "Measuring the Effectiveness of Crop Improvement Research in Sub-Saharan Africa from the Perspectives of Varietal Output, Adoption, and Change: 20 Crops, 30 Countries, and 1,150 Cultivars in Farmers' Fields." Standing Panel on Impact Assessment (SPIA), CGIAR (Consultative Group for International Agricultural Research Program on Policies, Institutions, and Markets) Science Council, Rome.

Ward, C., with R. Torquebiau and H. Xie. 2016. *Improved Agricultural Water Management for Africa's Drylands*. World Bank Studies. Washington, DC: World Bank.

You, L., S. Wood, and U. Wood-Sichra. 2009. "Generating Plausible Crop Distribution Maps for Sub-Saharan Africa Using a Spatially Disaggregated Data Fusion and Optimization Approach." *Agricultural System* 99(2–3): 126–40.

Xie, H., W. Anderson, N. Perez, C. Ringler, L. You, and N. Cenacchi. 2015. "Agricultural Water Management for the African Drylands South of the Sahara." Background report for the Africa Drylands Study. International Food Policy Research Institute, Washington, DC.

## Chapter 8

# Healthy Ecosystems: Integrated Approaches for Well-Balanced Landscapes

*Erin Gray, Norbert Henninger, Robert Winterbottom, Chris Reij, Paola Agostini*

## Current situation

In Sub-Saharan Africa as in other parts of the world, dryland communities, along with their production systems and human livelihood strategies, have evolved over hundreds of years in response to an unfavorable climate, enabling both ecosystems and human well-being to recover following droughts, floods, and fires. Over the past decades, however, high human population growth rates, increasing land use pressures and associated land degradation, changes in rainfall patterns, greater frequencies and intensities of droughts, intensifying conflicts over natural resources, and other natural and anthropogenic drivers have begun to undermine the resilience of many dryland communities in Africa and have contributed to depleted soil fertility and water stress. An increasing number of these communities are facing a reduced capacity of the land to support them, lowering their resilience to recover from natural shocks.

Although efforts to address these challenges in the drylands of Sub-Saharan Africa have yielded some positive outcomes, all too often they have failed to achieve significant and lasting improvements at scale. In an environment in which water is frequently the most limiting resource, few interventions have taken adequate account of linkages between upstream and downstream water users. In many cases, well-intentioned interventions have disrupted traditional management systems for common pool resources, such as wetlands, grazing reserves, and forests.

Single-objective and sectoral development approaches in particular are increasingly seen as inadequate because they may not fully address trade-offs associated with competing land uses and actors, or they fall short in incorporating the perspectives of all stakeholders in local communities and in appropriately addressing sources of resource conflict. They may also fail to take into account biophysical connections and leverage interactions among production systems that are critically important in dryland systems and necessary to generate and sustain both farm-level and landscape-level benefits. For example, trees in agricultural landscapes can play a critical role in renewing soil fertility, providing fodder for livestock, and generating fuelwood for households, while simultaneously contributing to the diversification and enhanced resilience of farming systems; yet many agricultural and livestock development programs have not taken fully into account the key roles of trees in agricultural landscapes.

Many development actors across Sub-Saharan Africa are starting to adjust dryland development programs so as to consider multiple objectives and multiple actors across two or more sectors. Evidence is emerging that a carefully sequenced landscape approach can increase the effectiveness of development programs and capitalize on opportunities to restore resilience in drylands.

## Opportunities

Water scarcity and land degradation are the major biophysical constraints facing drylands and are key threats to economic development and human welfare. Sustainable land management interventions that can conserve soil and water, build natural and social capital, and maximize efficiency of water and soil resource use can be critical for stabilizing rural production systems. They can also help rebuild household resilience. In many locations these interventions can be considered foundational for sustainable agricultural intensification. A number of practices have been identified as especially promising for drylands, where the need for the widespread adoption of improved land and water management practices to boost productivity is especially acute. These practices include agroforestry, farmer-led soil and water conservation techniques, rainwater harvesting, conservation agriculture, and integrated soil fertility management. These measures can be extremely effective in reversing land degradation and contributing to the sustainable intensification of agriculture and forestry. Rural economies can benefit from these practices through higher crop yields; increased supplies of fodder, firewood, and other valuable goods; greater income and employment opportunities; a restoration of biodiversity and ecosystem services; and enhanced resilience in the face of climate change. Promoting the widespread adoption of these improved land management practices can be a core element of integrated landscape management designed to enhance and diversify production systems and increase household resilience.

### Integrated landscape management

Integrated landscape management represents an opportunity to restore dryland areas in Sub-Saharan Africa. The definition of integrated landscape management adopted here is based on the definition used by members of the initiative on Landscapes for People, Food and Nature, a collaborative partnership of environmental and agricultural nongovernmental organizations (NGOs), UN agencies, and governments:

> *[Integrated landscape management is characterized by] long-term collaboration among different groups of land managers and stakeholders to achieve the multiple objectives required from the landscape. These typically include agricultural production, provision of ecosystem services (such as water flow regulation and quality, pollination, climate change mitigation and adaptation, cultural values); protection of biodiversity, landscape beauty, identity and recreation value; and local livelihoods, human health and well-being. Stakeholders seek to solve shared problems or capitalize on new opportunities that reduce trade-offs and strengthen synergies among different landscape objectives. Because landscapes are coupled socio-ecological systems, complexity and change are inherent properties that require management.* (Scherr, Shames, and Friedman 2013)

Integrated landscape management provides a framework for scaling and leveraging land and water management interventions in such a way that the whole becomes greater than the sum of individual interventions in terms of ecological and economic gains. Scherr, Shames, and Friedman (2013) identified five key actions for operationalizing integrated landscape management to promote successful dryland restoration and community resilience: (1) Interventions are designed to promote multiple goals and objectives; (2) Ecological, social, and economic interactions are managed to reduce negative trade-offs and optimize synergies; (3) Roles of local communities are acknowledged; (4) Planning and management of interventions is adaptive; and (5) Collaborative action and comprehensive stakeholder engagement are encouraged and institutionalized.

Following the Scherr, Shames, and Friedman (2013) report and taking into account the findings of an extensive literature review, this chapter categorizes the five key actions into three broad core components for operationalization, and provides 10 key principles that can be viewed as a checklist for implementation and operationalization of integrated landscape management (see figure 8.1). The three core components are as follows.

**Core Component 1: Landscape Goal(s) Encompassing Multiple Objectives at Different Scales.** In drylands with mixed land uses and multiple stakeholders, it is important to establish a shared perception of dryland landscapes by identifying and fostering multiple objectives and goals. This promotes common entry points among stakeholders for collaboration around actions that are critical to enhancing resilience. In the drylands of Sub-Saharan Africa, goals and objectives generally relate to improving food security and livelihood diversification. In some higher-producing regions, goals often also include sustainable intensification of production systems. Integrated landscape management must consider multiple scales upon which to implement interventions (e.g., farm level and landscape level), as well as temporal and biophysical dimensions—which are especially important in environments with highly variable and unevenly distributed rainfall.

Integrated landscape management must generate short-term economic returns to incentivize farmer and herder participation, but it must also promote thinking holistically about maximizing ecological gains, for example, by improving biophysical connectivity to restore groundwater levels, by providing critical corridors for livestock movement, or by preserving wildlife habitat. Integrated landscape management must also consider the multi-functionality of landscapes and provide a mechanism that enables local stakeholders to reduce conflicts among different types of specialized resource users who differ in their dependencies on a range of ecosystem services (e.g., herders, farmers, or fishers).

**Core Component 2: Adaptive Planning and Management.** Integrated landscape management must seek to understand how land users interact with their environment and take advantage of key sources of income that can improve welfare. The planning of land use, grazing, and natural resource use under integrated landscape management must recognize ecological, social, and economic interactions among different parts of a landscape, which then can be managed to optimize synergies and reduce negative trade-offs. Integrated landscape management should promote continual learning from outcomes and create opportunities to scale up successes and address failures. Adaptive management is also important for understanding the resilience of a landscape—for example, how it responds to shocks such as changes in rainfall or temperature. As climatic and economic risks create uncertainty, adaptive planning and management—whereby stakeholders review at recurring intervals the successes and challenges of current land use choices—allows all involved to quickly address risks. As such, integrated landscape management requires effective, user-friendly, participatory monitoring and evaluation systems and feedback mechanisms.

**Core Component 3: Collaborative Action and Comprehensive Stakeholder Involvement.** Integrated landscape management must recognize the critical importance of identifying and acknowledging the roles of local communities and households in resource management. Integrated landscape management must promote community-wide participation and planning in dryland restoration and other land use interventions, collective action for implementing these interventions, and coordination among key stakeholders across scales and sectors. For example, on steep slopes, actions taken by farmers to minimize tillage operations combined with actions taken by herders to reduce grazing pressure in critical locations will have greater impact on erosion and sedimentation rates and restoration of the vegetative cover than fragmented or individual efforts alone. Local communities must be incentivized to invest in improved land and water management and to share local knowledge and experience.

Based on the three core components, 10 key principles are identified for integrated landscape management (figure 8.1). The 10 key principles are useful for design processes that can motivate multiple stakeholders to pursue a set of common goals within a landscape, explicitly recognize synergies and trade-offs between different objectives, and establish agreed mechanisms to resolve differences.

Skeptics of integrated landscape management may characterize the quest for enhanced integration across multiple sectors and stakeholders, along with greater emphasis on geographic targeting, as nothing new. Conceptually, however, the way integrated landscape management is being proposed here is novel in that it incorporates lessons learned from previous land management approaches and places much greater emphasis on building resilience to drivers such as climate change and changing market forces. Integrated landscape management provides added value in that it:

- Does not promote a "one-size-fits-all" approach but rather asks stakeholders to consider the local context and take sectors, stakeholders, and social, cultural, and other conditions into account across geographic boundaries that make ecological sense. Integrated landscape management promotes a flexible framework for scaling investments at a landscape scale to maximize ecological, economic, and social synergies and minimize negative trade-offs.

**Figure 8.1** Core components of integrated landscape management

*Source:* Sayer et al. 2013.

- Emphasizes that planning and implementation take into account spatial components important to rejuvenating and maintaining ecosystem health (e.g., hydrological flows, habitat). Integrated landscape management requires that land use planners and decision makers think differently about scale and take into account these spatial components.

- Promotes a combination of bottom-up and top-down principles designed to encourage local community participation, but at the same time remains committed to building appropriate institutional and financial support.

- Promotes an adaptive management approach that tries to build in monitoring and evaluation to generate long-term data needed to truly understand whether communities are becoming more resilient and increasing their adaptive capacity, and whether landscape-level changes are achieved.

Integrated landscape management has the potential to improve dryland development efforts by bringing about the following intermediate results:

- **Increased action and investment from stakeholders.** Community-driven integrated restoration of small watersheds in Tigray, Ethiopia, for example, has motivated farmers to invest in improved soil and water management practices. Their coordinated efforts have resulted in recharged groundwater tables in valley bottoms, allowing farmers to develop dry season irrigation and cultivate higher-value crops.

- **Reduced conflict over use of land and other resources.** Improved coordination among stakeholders can help clarify rights and responsibilities and improve understanding of landscape goals and objectives. This in turn can lead to reduced conflict over land and other resources. Good examples are the agreements negotiated between farmers and herders to demarcate corridors for livestock movement, which have helped to protect farmers' crops and trees from livestock browsing while at the same time safeguarding grazing and water access areas for herders; and the agreements negotiated between local communities and firewood and charcoal merchants, which have helped merchants source wood from locally managed forests and tree farms while at the same time contributing to sustainable management by the local communities of what has become an important livelihood source.

- **Economies of scope and scale.** Land and water users in a landscape sharing their skills and assets can achieve economies of scale and exploit cost advantages resulting from integrated production. Some landscape interventions also lead to increases in household income associated with the simultaneous production of two or more products.

- **Capacity building.** Integrated landscape management promotes community participation and collective action, so that farmers, herders, and other resource users learn about new sustainable practices and technologies. Local institutions are empowered to negotiate and adopt rules to improve environmental governance, provide for more equitable benefit sharing, and accelerate the adoption of improved natural resource management practices.

- **Resilience at the household and landscape level.** Collective action by a large number of households can affect all three dimensions of resilience, depending on the local circumstances: *exposure to shocks* (e.g., households in southern Niger reported experiencing reduced wind speeds at the beginning of the growing season after they increased on-farm tree densities); *coping capacity* (e.g., dryland farmers in the Kitengela Plains of Kenya benefited from new income sources after they were persuaded to leave wildlife migration routes unfenced, improving wildlife and tourism benefits for nearby Nairobi National Park); and *sensitivity to shocks* (e.g., households in Tanzania diversified their livelihood strategies and were able to buffer dry season risks for livestock after they restored woodlands and expanded dry season grazing areas through assisted natural tree regeneration).

## Benefits of integrated landscape management

Integrated landscape management for drylands typically revolves around reversing land degradation and improving ecosystem health and functionality. As such, the benefits of integrated landscape management are intricately tied to the ability of ecosystems within a target area to generate services. Dryland ecosystems are able to provide a variety of economically valuable goods and services. IUCN-ESARO (2010) divides the range of ecosystem services provided by drylands into four categories: (1) cultural services, (2) provisioning services, (3) regulating services, and (4) supporting services. Examples are depicted in table 8.1.

**Table 8.1** Ecosystem services provided by drylands in Africa

| Cultural | Regulating | Provisioning | Supporting |
|---|---|---|---|
| • Recreation and tourism | • Micro-climate regulation and carbon sequestration<br>• Pollination and seed dispersal<br>• Water and air filtration/purification<br>• Erosion control | • Food and honey<br>• Fodder<br>• Timber and non-timber forest products<br>• Freshwater<br>• Energy<br>• Medicinal and cosmetic products<br>• Habitat | • Soil development<br>• Nutrient cycling<br>• Primary production |

*Source:* IUCN-ESARO 2010.

*Note:* Biodiversity in drylands provides the foundation for all four types of ecosystem services. Biodiversity is generally not defined as an ecosystem service.

These benefits can be categorized as market and non-market benefits. Cultural services are generally non-material, but they can add to well-being (e.g., tourism). Cultural services can generate economic value, which can be captured and converted to income for the benefit of local communities (e.g., entrance fees to national parks). Regulating services are benefits generated by an ecosystem's ability to regulate natural processes (e.g., air and water filtration). Regulating services can be more difficult to quantify, especially if biophysical information on these processes that links them to human welfare is lacking, and often market prices are not available for economic valuation. Provisioning services are benefits that people can directly extract from ecosystems (e.g., support to farming and herding). Many provisioning services are easily valued, as market prices are readily available for things like crops and animals. However, some provisioning services generate non-market benefits that are difficult to value (e.g., maintenance of biodiversity). Supporting services are those that underlie provisioning and regulating services, and as so are generally not valued in an economic analysis.

Beyond enhancing the provision of ecosystem services, integrated landscape management provides social benefits related to investments in social or human capital, health, and improved access to resources and markets. Many landscape management interventions focus on building community-level institutions, such as farmer cooperatives or local savings and loan associations. This building of social capital generates multiple market and non-market benefits, as it serves to diversify income, improve education and equality, and spread awareness of the value of sustainable land management, which can help to reduce degradation in the future.

An important mechanism through which integrated landscape management delivers social benefits is by fostering collective action. Land management practiced at landscape scale can deliver greater benefits than when practiced at farm scale because collective action can:

- Allow resource users to more easily manage ecosystems across geographical, cultural, and political boundaries.

- Increase uptake of sustainable land-use practices, as resource users are more likely to adopt practices if they see their neighbors conducting and benefiting from these practices.

- Make it easier for resource users with different skillsets to collect, share, and create knowledge, skills, and assets at a lower cost.

- Encourage communication and coordination among diverse interest groups and stakeholders, reducing conflict over natural resources that can result in violence, land degradation, and project disruption. Collective action can improve communication between resource users, reducing the costs of conflict resolution around local issues.

Collective action can result in economies of scale and scope, reducing transaction and implementation costs and enhancing benefits. Based on a survey of collective action institutions found in East Africa, Mogoi et al. (2009) identified a wide range of benefits of integrated landscape management (see table 8.2).

## Challenges

Several barriers need to be overcome before integrated landscape management can become part of regular policy making and development planning in Africa's drylands:

- **Lack of knowledge and awareness about integrated landscape management among national and local governments, the private sector, and civil society actors.** Landscape-level ideas have yet to percolate down to more national and local actors. One reason is that many integrated landscape management programs lack strong monitoring and evaluation components, especially beyond a household and community scale, making the assessment of landscape-level benefits difficult.

- **Institutional barriers that impede addressing complexities at the landscape level.** Landscape dynamics are usually very complex, as they involve interactions between diverse groups of stakeholders and different land uses. In most

**Table 8.2** Benefits of integrated landscape management

| Market benefits |
| --- |
| • Improved agricultural, forestry, fuelwood, fodder productivity |
| • Carbon sequestration |
| • Avoided transaction costs |
| • Avoided siltation and flooding costs |
| • Water quality and quantity regulation |
| • Pollination services |

| Non-market benefits |
| --- |
| • Avoided costs of travel time to procure water, fuelwood, and other supplies |
| • Avoided costs of conflict |
| • Female empowerment |
| • Increases in biodiversity and improved habitat |
| • More opportunities for recreation |
| • Increases in traditional knowledge |
| • Improved access to health services, markets, and education |
| • Improved resilience (e.g., avoided costs from drought) |
| • Stronger cultural values |

*Source:* Mogoi et al. 2009.

cases, there are no simple solutions to complex challenges, and "one size fits all" approaches do not work. A careful assessment of location-specific challenges is required, as well as a learning-by-doing approach combined with a significant investment in institutional reforms and capacity building. In addition, ways need to be found to take into account sector-specific mandates of different ministries to resolve the challenges of working across sectors.

- **Poor availability of and access to location-specific data about land, water, and natural resource use.** For many dryland areas, local planners have very limited access to geographic information system (GIS) data relating to land cover, land use, water supply, water abstraction, and other natural resource uses. This may be because the data do not exist or are not publicly available. Without detailed and reliable data, it is difficult to develop effective landscape management strategies.

- **Difficulty in ensuring management of trade-offs and provision of adequate incentives for needed behavior changes and sustainability.** In the complex agro-pastoral production systems found throughout many parts of the drylands, there is a special need to assess trade-offs and synergies between different land uses and users. However, capacity for this type of analysis among implementing agencies is generally weak.

- **Fragmented financing and planning for dryland restoration to optimize land use.** Local land use planning capacity is generally low because of a persistent marginalization of drylands. Failure to address local land use planning can result in conflict over resources and land areas and other costs.

## Key messages

By providing a comprehensive framework that can be used to exploit synergies among a wide range of more focused interventions, integrated landscape management can help reverse land degradation and improve ecosystem health and functionality in the drylands of Sub-Saharan Africa. Increased investment in integrated landscape management programs, which support coordination and long-term collaboration among different groups of land managers and stakeholders within dryland landscapes, can enhance and safeguard land restoration efforts, lower risks related to water shortages and soil fertility declines, allow local populations to raise their incomes and diversify their livelihood sources, support sustainable intensification, and reduce conflicts. In this way, integrated landscape management can potentially serve as the unifying framework for efforts to enhance the resilience of vulnerable populations in the drylands.

# References

IUCN-ESARO (International Union for Conservation of Nature, Eastern and Southern Africa Regional Office). 2010. "Drylands Situation Analysis." https://cmsdata.iucn. org/downloads/iucn_esaro_drylands_situation_analysis.pdf.

Mogoi, J., J. Tanui, W. Mazengia, and C. Lyamchai. 2009. "Role of Collective Action and Policy Options for Fostering Participation in Natural Resource Management." Paper presented at 2nd World Congress on Agro-forestry, August 24–28. Nairobi, Kenya.

Sayer, J., T. Sunderland, J. Ghazoul, J.L. Pfund, D. Sheil, E. Meijaard, M. Venter, A. Klintuni Boedhihartono, M. Day, C. Garcia, C. van Oosten, and L.E. Buck. 2013. "Ten Principles for a Landscape Approach to Reconciling Agriculture, Conservation, and Other Competing Land Uses." *Proceedings of the National Academy of Science* 110 (21): 8349–8356. Accessed on 28 February 2014 at http://www.pnas.org/content/ early/2013/05/14/1210595110.full.pdf+html

Scherr, S.J., S. Shames, and R. Friedman. 2013. "Defining Integrated Landscape Management for Policy Makers." EcoAgriculture Policy Focus No. 10. EcoAgriculture Partners, Washington, DC.

## Chapter 9

# Market Connections: Promoting Trade to Promote Resilience

*John Nash, Paul Brenton, Alvaro Federico Barra*

## Current situation

Good trade policy is a crucial ingredient to economic development worldwide, but there are reasons to believe that it is especially important in dryland regions of Africa. Enhanced trade could contribute to reduced vulnerability and increased resilience of poor households living in the drylands in at least three ways.

### Gains in agricultural productivity

First, enhanced trade could help drive productivity gains in agriculture. Agricultural productivity in the drylands is already low compared to other developing regions, and in the future the gap could widen as a result of global warming. Enhanced flows of technology are critical for improving productivity and adapting to a changing climate. Technology embodied in imported inputs—for example, seed of improved crop varieties, fertilizer, agricultural machinery, and animal vaccines—would pave the way for the emergence of more intensive production systems with increased productivity and greater sustainability (Jouanjean 2013).

In the drylands trade barriers now impede adoption of improved production technologies. Africa as a whole has fertilizer usage rates far below those seen in other regions, in part because fertilizer prices in many parts of Africa are some of the highest in the world. Trade barriers, both official (e.g., tariffs) and indirect (e.g., regulations), keep prices high and discourage companies from entering African markets. They also impede market integration, keeping markets small and preventing the realization of economies of scale in manufacturing and importation of fertilizers, which could help reduce prices. In seed markets African farmers (outside of South Africa) typically have available an average of less than one new variety of maize a year, far less than in other countries outside the region and far below the number that would be needed to trigger transformational change (Gisselquist et al. 2013).

### Positive effects on food prices

Second, enhanced trade could help increase food prices received by producers, reduce food prices paid by consumers, and dampen food price variability for both groups. Urban and rural poverty rates are high in dryland areas, which means that food prices affect many people both in cities (where higher food prices punish consumers, disproportionately affecting the poor) and in rural areas (where lower prices received for sales of agricultural commodities undermine the income of producers). Trade barriers that increase transaction costs exacerbate both problems. Improved trade can reduce the wedge between producer and consumer prices, increasing the welfare of consumers in structural deficit areas where food prices are high and of producers in surplus areas where farm gate prices are relatively low. For example, USAID (2011) estimates that in West African cereals markets a reduction in transaction costs equivalent to 10 percent of the farm gate price could stimulate a 4 percent increase in production and a proportionally similar increase in the real incomes of farmers, while at the same time causing an 8 percent reduction in real consumer food prices.

In dryland areas, which are particularly vulnerable to both climatic and manmade disasters and associated food production shocks, increased integration with larger regional markets can reduce the magnitude of the price effects from localized shocks, while lower barriers and better trade infrastructure can allow faster and more efficient response to localized food shortages due to disasters of all types. Badiane, Odjo, and Jemaneh (2014) examined food production variability in the countries of the Economic Community of West African States (ECOWAS), among others, finding that in every country volatility (measured as the coefficient of variation) of within-country production is higher than that of the ECOWAS region as a whole, and that production is imperfectly correlated across countries. The clear implication is that greater intra-regional trade, even within ECOWAS, would help stabilize prices in individual countries in the face of local shocks. This would hold *a fortiori* for the larger pan-African regional market.

Currently, food markets in many dryland areas remain fragmented, isolated from regional and global markets. Haggblade (2013) cites numerous examples throughout Africa of food surplus areas that are separated from nearby deficit areas by political boundaries, which artificially divide these natural "market sheds." In dryland regions, examples include surplus millet- and sorghum-producing areas in Mali and Burkina Faso that are separated from natural markets in a half dozen neighboring countries, as well as livestock producing regions in Mali, Mauritania, and Niger that are separated from natural markets in nearby coastal countries. Evidence is strong that borders do indeed have a substantial negative impact on trade (map 9.1). In West Africa cereal prices differ dramatically between net producing and net consuming markets, providing evidence of this lack of integration. One way of quantifying this effect is to evaluate what would be the increased distance that would increase costs equivalently to crossing a border. Analysis of prices of

**Map 9.1** Maize hotspots in drylands regions where production may be discouraged by trade barriers

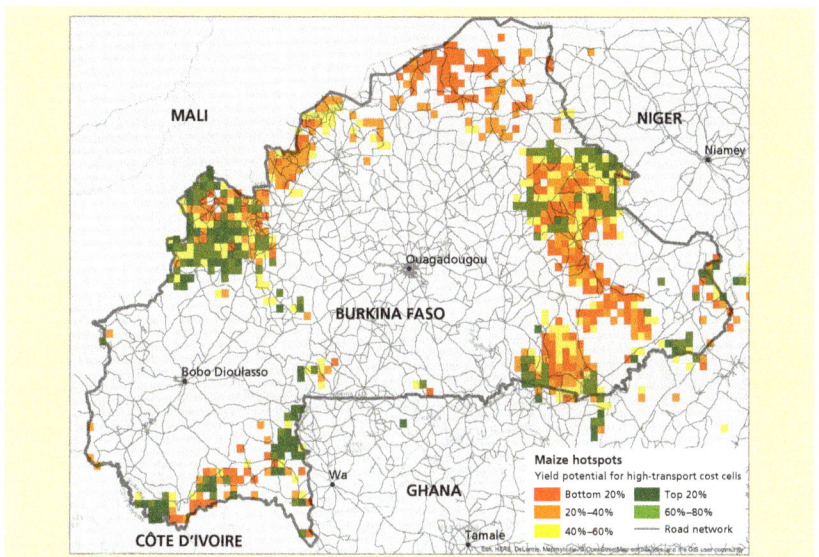

*Source:* World Bank.

maize, rice, and cowpeas by Brenton, Portugal-Perez, and Regolo (2014) found that crossing the border between Niger and Nigeria is equivalent to pushing these countries 639 kilometers (km) further apart; the Nigeria-Chad border effect is equivalent to adding 594 km.

Further evidence of the effects of poor integration is found in the volatility of relative prices of food staples in neighboring markets, which is much higher when the markets are located on opposite sides of a border than when they are located within the same country. Figure 9.1 shows the volatility of monthly prices of millet, with much greater dispersion between markets in different countries than between markets within a country. While some of the volatility is due to changes in the cost of transporting goods between countries, a significant portion is attributable to other costs associated with crossing borders. As an example, crossing the border between Ghana and Togo appears to increase the volatility of food staple prices by over 40 percent compared to markets within each of the two countries, suggesting a very low level of trade integration between these countries.

## Stimulate business development and employment

Third, enhanced trade could help stimulate business development and boost employment. Facilitation of trade in crops, livestock, and inputs brings the prospect of a significant number of new jobs. A vigorous agriculture and agribusiness

**Figure 9.1** Relative prices of millet in West African markets, 2007–13 (in logarithms)

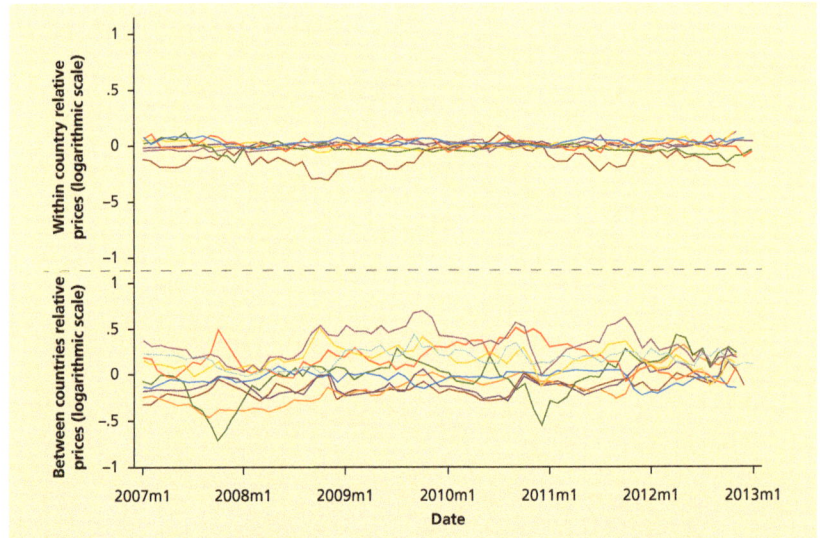

*Source:* Brunelin and Portugal-Perez 2013.

*Note:* Each line in the figure shows the relative retail millet prices (on a monthly basis) between two markets, either within the same country (upper part of figure) or between two countries (lower part of figure). The sample includes 173 markets located in 14 West and Central African countries.

sector creates jobs in activities all along the value chain—in producing and distributing seeds and fertilizers; in providing extension advisory services; in assembling, processing, and storing grains; and in transporting, distributing, wholesaling, and retailing agricultural products (Brooks, Zoriya, and Gautam 2012).

## Opportunities

When food markets are not working properly, grain storage strategies are sometimes used in an effort to limit exposure to variability in food prices (box 9.1). But efforts to improve market integration can contribute to the resilience of households living in drylands. To see how, it is useful to consider how enhanced trade might affect the three determinants of resilience.

### Reducing exposure

To the extent that enhanced trade can reduce the frequency and dampen the severity of food price spikes, poor households living in drylands that rely on markets to meet some or all of their consumption requirements would be less exposed to economic shocks.

BOX 9.1

## Limiting exposure to price shocks through grain storage strategies

When food markets are not working properly, grain storage strategies are sometimes used in an effort to limit exposure to variability in food prices. Such strategies may be implemented at the household, community, or national level.

**Household level.** Farmers sometimes try to protect themselves against seasonal food price variability by storing a portion of what they produce to bridge the time between harvests. Doing so insulates them against seasonal price swings, but it comes at a cost. When crops are stored in traditional bins, typically a portion is lost to rodents, insects, mildew, and/or theft. The losses can be significant. For example, in the case of maize, Tefera (2012) estimates that African farmers lose 4–10 percent of what they store. Improved storage technologies—for example reinforced polyurethane bags or small-scale metal silos—can be effective in limiting on-farm storage losses (Gitonga et al. 2013), but these technologies are not always cost-effective.

**Community level.** As food markets become more integrated, private companies often emerge to offer storage services. By operating at a larger scale with better technologies, they may be able to reduce the cost of storage losses while at the same time allowing farmers to hold onto supplies in anticipation of higher prices later in the season. This in turn helps smooth out seasonal price swings. Since the most cost-effective way of storing grain is to keep the grain loose and unbagged, grain from different owners is mixed, so mechanisms have to be in place to make sure that the grain farmers take out of storage is equal in quality to the grain they put in, and that they are compensated if there is a deterioration in quality. In some countries private or public entities now offer the insurance, bonding, and quality inspection services needed to back private silo operators.

Grain held in certified warehouses can also be used as collateral for credit. In a number of African countries warehouse receipts—basically certificates showing ownership of stocks held in authorized warehouses—can now be sold and bought. Farmers can receive "loans" by selling their warehouse receipts while entering into a repurchase agreement with the warehouse or silo to buy back their receipts at a price based on prevailing interest rates. Futures contracts, which allow market participants to hedge or speculate on commodity prices, are often settled using warehouse receipts as well (Giovannucci, Varangis, and Larson 2000).

**National level.** Governments sometimes try to use grain storage policies to smooth food prices. Potentially, with the right set of rules, grain importing countries can build up large stockpiles and then release inventories to

*(continued next page)*

**Box 9.1** *(continued)*

dampen price spikes (Larson et al. 2014). Because the costs of administering such programs are usually quite high, this type of price stabilization tends to be more expensive than other measures, such as targeted safety nets. Importantly, a series of adverse effects can cause public stocks to run out, in which case the stabilization schemes are likely to fail just when they are most needed.

In recent years many African governments have revisited policies that use inventories or trade rules to manage high prices. During the 2008 global food crisis when grain prices spiked dramatically, the policies that seemed to make sense for individual countries led collectively to more volatile prices internationally (Martin and Anderson 2012). Recent experience also reconfirms how hard it is to anticipate market outcomes. For example, in Malawi recent research suggests that the government's well-intentioned efforts to manage food prices had the unintended effect of increasing price volatility (Ellis and Manda 2012).

**Regional trade liberalization.** In both East and West Africa, efforts are under way to promote market integration at the regional level. The East African Community (EAC) and ECOWAS are playing a leading role in these efforts. The Regional Economic Commissions (RECs) are making efforts to forge regional markets by reducing formal barriers and lowering technical barriers to trade by harmonizing standards and regulations. At the same time, individual countries can also act on their own when regional efforts bog down. Initiatives undertaken by subregional coalitions of willing members to fast-track implementation of agreements can in some cases be a more expeditious mechanism to enhance trade among participating countries. And in cases when harmonization of regulations and standards at the regional level lags, mutual recognition of standards and regulatory approvals among a subgroup of countries can provide benefits more quickly.

**Reducing or eliminating non-tariff measures.** Non-tariff measures (NTMs) represent a major obstacle to enhanced trade in food and food products in the drylands, restricting the availability of food in the market and driving a wedge between producer and consumer prices. In West Africa a number of nongovernmental and advocacy organizations engage on agricultural trade policy (Pannhausen and Untied 2010). In the ECOWAS region an effort is under way to combat the lack of private sector awareness of regional protocols by inviting nonstate actors to participate more actively in regional forums on the implementation of Economic Community of West Africa Agricultural Policy (ECOWAP) (see Harris, Chambers, and Foresti 2011). More generally, however, policy making is often still driven by REC authorities, donor agencies, and national governments, with other actors exerting little influence. Lack of information is also an issue: there tends to be limited awareness of the scope and nature of NTMs. ECOWAS has put in place national committees to deal with problems caused by NTMs and has established complaint desks at borders, but it remains unclear whether these measures are having a significant impact. In East Africa a different approach is being used: the Common Market for Eastern and Southern Africa (COMESA), EAC, and the Southern African Development Community (SADC) have set up an online database that seems to have been effective in focusing policy makers' attention on the problem (see www.tradebarriers.org). What seems clear is that if NTMs are to be reduced, governments must ensure that all rules and regulations affecting regional trade in food and crop inputs are clearly available at the border, well-known by traders and officials, and applied in a consistent manner.

## Reducing sensitivity

To the extent that enhanced trade can reduce the cost of using improved production technology, poor households in drylands that are dependent on herding and farming could improve and stabilize their incomes and thereby reduce their sensitivity to weather- and disease-related shocks.

**Integrate input markets to facilitate technology flows.** Improved technology such as that embedded in new crop varieties, new breeds of animals, new fertilizer types, new crop chemicals, and new types of machinery can reach farmers and herders in two ways: the technology can be imported or it can be developed domestically. Both channels are discouraged by high regulatory costs, particularly costs associated with mandatory performance testing. Faced with onerous requirements to put new products through long and expensive tests to prove that they work well, companies will choose not to enter the small markets represented by most African countries. In many dryland countries legal restrictions make it difficult or impossible to distribute improved crop varieties or animal breeds without first carrying out expensive and time-consuming testing, even if the varieties or breeds have been imported from a neighboring country with similar agro-ecological conditions. Since the risks associated with release of ineffective technologies are low, regulatory policy reforms are needed to accelerate the introduction of new crop varieties and animal breeds, along with other innovative production technologies. Measures to achieve this could include: (1) eliminating or at a minimum streamlining performance testing (which experience in other countries has shown can be left to the market, with increased enforcement of anti-fraud and truth in labeling laws being the focus of government agencies); (2) making certification optional, as it is in South Africa, Turkey, and the United States, among other countries; and (3) relaxing restrictions on international trade.

As discussed earlier, enhanced trade in improved plant varieties offers particularly attractive prospects in the drylands for improving the resilience of households that depend on agriculture as a primary livelihood activity. The flow of new plant varieties from other countries could be facilitated by regional harmonization of varietal testing and release requirements, which could be mediated through one of the RECs. One approach would be to establish regional seed catalogues with the understanding that any variety entered in the regional catalogue could be used throughout the region without further registration requirements. Harmonization would create much larger markets for inputs, which could be more attractive for private trade and investment. In West Africa the members of ECOWAS have agreed in principle to harmonize regulations for plant varieties, but progress in implementing the agreement has been slow. Many of the benefits of regional harmonization could be achieved more quickly if individual countries were to agree to accept varieties that have already been tested and approved for release in other countries ("mutual recognition"). Unilateral and bilateral agreements of this type could be made consistent with the full regional agreement once the latter is agreed and implemented. A similar approach is being used by the European Union, where varieties approved by any one member state are automatically approved for use in all other member states.

Lowering of trade barriers similarly could expand markets for fertilizer. Many countries in East and West Africa continue to maintain restrictions on international trade in fertilizer. The Abuja Declaration on Fertilizer for the African Green Revolution endorsed by African Union Ministers of Agriculture on June 12, 2006 calls for removal of duties on imported fertilizer. Equally important is the elimination of regulatory barriers to international trade in fertilizer, such as regulations that each blend must be tested and approved for use in every country. These barriers to trade can be reduced by approving fertilizer ingredients, rather than finished fertilizer products, and by automatically recognizing ingredients approved in neighboring countries.

### Improving coping capacity

To the extent that enhanced trade can improve the performance of food markets after a shock has hit, poor households living in drylands that are dependent on herding and farming will have an easier time coping with the effects of the shock.

**Limit the use of trade barriers to cope with temporary food price spikes and localized production shortfalls.** In the drylands of Africa as elsewhere, governments have tended to respond to food price spikes by enacting pro-cyclical trade policies, such as temporary reductions in import protection or temporary increases in export barriers during periods of high food prices. A growing body of evidence makes clear that such policies have likely amplified food price volatility at the global level (for examples, see Anderson, Ivanic, and Martin 2013; Headey 2010; Karapinar and Haberli 2010; Mitra and Josling 2009; Martin and Anderson 2012; Magrini et al. 2013; and Rutten, Shutes, and Meijerink

2011). In the face of domestic pressure, it is often difficult for governments to make a credible commitment to refrain from such policies when food prices spike. But in addition to their collective global impacts, these ad hoc interventions also have perverse local consequences because they discourage investment in trade activities and infrastructure. As a result, local price volatility would likely be lower without them. In a study of food price volatility in Africa, Minot (2012) notes that, "These findings are consistent with a number of studies that suggest that unpredictable government intervention in maize markets, and the trade restrictions that often accompany these policies, can inhibit private traders from participating in trade and storage activities, thereby increasing seasonal volatility and exacerbating price spikes associated with supply shortfalls." The findings of Magrini et al. (2013) underscore this lesson. Using propensity score matching to control for selection bias, they find that countries that rely on trade distortions are more vulnerable to food insecurity, as measured by food availability. Similarly, Brenton, Portugal-Perez, and Regolo (2014) show that countries in Africa with less-integrated domestic markets and thicker borders have a higher prevalence of food insufficiency.

Trade barriers are of particular importance in the dryland areas of Africa, which are subject to high variability in rainfall and other climatic conditions. The highly variable rainfall leads to significant risks to those dependent on agriculture and animal husbandry, and it is the reason why a large proportion of farmers and herders in the drylands of Africa are small-scale, resource-poor, and subsistence-based. Trade policy in these regions should be carefully designed and implemented to assist farmers and herders—especially small farmers who do not now have access to mitigating strategies such as diverse crops or market mechanisms such as weather insurance—to cope with the natural variation in the weather, rather than increasing the unpredictability and uncertainty surrounding agricultural markets and trade. Instruments other than trade policies are available to meet the objective of reducing price volatility; for example, allowing governments or private traders to put a ceiling on future import prices. Futures contracts and call options have been available on the South African Futures Exchange (Safex) since 1999 (Haggblade 2013). The World Bank helped arrange a contract with these instruments for the government of Malawi, with the premium financed mainly by donors (Slater and Dana 2006). This could be a model worth exploring with other countries that would agree to commit not to use ad hoc trade barriers.

To make matters worse, ad hoc trade policies often are deployed not because a crisis exists, but simply out of fear that a crisis might be coming. For this reason, improvements are needed in national data systems so that governments can have access to the information needed to make decisions based on concrete evidence and transparent rules. Even though significant efforts have been made in recent years to improve the collection and quality of data on prices, food production, and other key indicators, many countries still face substantial data gaps that limit the ability to devise evidence-based trade and agricultural policies, especially in crisis situations. The results have been very ad hoc and unpredictable policy making (Jayne and Tschirley 2009).

Better data collection at the national level, including data on stocks, and the transparent dissemination of data would help break the vicious cycle in which governments intervene in food markets when they believe private traders are not holding adequate stocks and would be unwilling or unable to import food to make up the deficit, and private traders refrain from holding stocks and from importing for fear they will be forced to sell at a loss due to government policy. Current policies often result in the worst of all worlds—extensive government participation in food markets, combined with high price volatility (Jayne and Tschirley 2009; Minot 2012).

Improved agricultural market information can help reduce the damage caused by inappropriate trade policies. Badu (2013) highlights the positive impacts of the Famine Early Warning Systems Network (FEWS NET), which has made food price data continuously available to policy makers in many southern African countries through monthly newsletters. For market information systems to be effective, however, they must be trustworthy, and for that reason it can be important for outside agencies to conduct assessments jointly with government departments, as some policy makers lack confidence in externally driven monitoring systems.

Information on food stocks can be especially important in alerting market participants in both the public and private sectors to impending crisis (Wiggins and Keats 2013). When reliable information is available on a timely basis, government authorities generally are more willing to commit to following transparent rules, which in turn can provide farmers and private traders greater certainty regarding the basis on which to make long-term investment decisions.

## Challenges

In the drylands of Africa as elsewhere, the main challenges associated with trade policy reform are political, as moves to liberalize trade are likely to have negative consequences for some vested interest groups. Commitments are often made at a regional level to lower trade barriers, but policy reforms agreed at regional level must be implemented at country level, and frequently that is where things bog down. The frequent failures to implement trade reforms are not accidental; rather they are the outcome of domestic political processes, as the groups that benefit from the status quo often have the power to resist change. Trade reforms designed to reduce the gap between producer and consumer prices have the potential to benefit farmers and poor consumers, but intermediaries earning rents—both in public agencies and in established private firms—stand to lose. Opening trucking markets to greater competition has the potential to reduce marketing margins to the benefit of producers and consumers, but it will also reduce the oligopoly rents being made by incumbent firms.[1] Reducing testing requirements for inputs will increase the availability of new cultivars for farmers, but it will reduce the role played by national research bureaucracies, potentially reducing their rents, financial or non-pecuniary. To facilitate the process of policy reform, one solution is to provide compensation for the losers, either financial (in the form of payments) or non-financial (in the form of retraining and alternative employment).

The political dynamics undermining trade policy reform are frequently exacerbated by a lack of resources. Many governments do not have separate budgets to support activities and programs relating to regional integration. For many politicians and civil servants, policy reform is an ad hoc activity, and they will allocate resources only when a request is made or political pressure is applied (AfDB 2013).

The usual shortcomings of the political process are particularly acute when it comes to policies relating to pastoralism. Policy makers generally lack understanding of pastoral production systems and do not recognize the economic importance of informal cross-border trade, especially for these populations. In the case of livestock trade in the Horn of Africa, Aklilu et al. (2013) argue that this is the result of a systematic bias, as policy makers tend to come from highland regions and prioritize these agricultural regions over the drier pastoral lowlands that tend to rely on livestock-keeping. This results in their treatment of "the activity as economically marginal and illegal, often resulting in the random and punitive enforcement on traders and producers alike, including confiscating livestock and food products from merchants" (Aklilu et al. 2013). Enhanced efforts to educate influential decision makers about the important functions of pastoralist systems—economic, social, and environmental—may have a high payoff.

An additional challenge associated with improving the performance of markets in the drylands is the relatively high financial cost of infrastructure in the context of tight budgets. Investments to densify rural road networks in areas of high productive potential, but currently low connectivity, as well as investments to improve roads connecting net supplier and net demander areas, could significantly contribute to enhanced resilience. Of course, all such investments would need to be guided by a realistic evaluation of costs and benefits, also taking into account environmental and social costs. Spatial analysis can be informative in this evaluation. To generate the largest possible returns, it is important to prioritize investments on the basis of the best analysis available, and spatial analysis is useful as one element in this process.

## Key messages

The potential to develop well-integrated and competitive regional food markets in dryland regions of Africa is being thwarted by barriers to trade. Barriers fragment markets, raising food costs in structural deficit areas, lowering producer prices in structural surplus areas, and magnifying the effects of local supply shocks on prices. During food price spikes, ad hoc policy responses and counter-responses have often been used in an effort to control prices, but ad hoc responses tend to have adverse long-run consequences on food security, as they discourage private sector arbitrage trade and investments (e.g., in storage) that could help mitigate future price swings. This is particularly detrimental to poor consumers in chronic food deficit areas.

Improving productivity is critical to the future of agriculture in the drylands. African agriculture in general has on average the lowest rate of input use, lowest productivity, and highest yield gaps of any region in the world, and these trends are particularly pronounced in dryland areas. The productivity gap will become an even greater problem in the future, because advances in productive technologies will be increasingly focused on instilling resilience to climatic change, so lags in adoption will cost farmers their resilience to these shocks. Many factors contribute to the low rate of adoption of improved technologies in the drylands, but one important factor is the high cost coupled with limited availability of inputs that embody these technologies, a situation greatly exacerbated by direct and indirect trade barriers.

A number of ongoing initiatives are seeking to reduce barriers to trade in food and agricultural inputs. To succeed, these initiatives will have to overcome political resistance, as well as entrenched attitudes of mistrust between the government and trade communities. More transparent and better information for civil society on the presence and effects of trade barriers, and for government on the realities in local food markets, may facilitate reforms. A better understanding of the political economy may also help, and efforts to study this topic are ongoing.

When regional processes to reduce trade barriers prove slow and cumbersome, countries should not hesitate to explore bilateral and plurilateral paths to reform, which in turn can demonstrate the benefits from joint policy reform and coordination and so help invigorate broader regional integration.

Transport costs are currently very high in the drylands and need to be brought down, both through regulatory reform to increase competition, as well as by appropriate investments. Spatial analysis can help identify areas of high payoff for investments.

Finally, policy makers need to find ways to take advantage of the informal trading systems that currently exist in the drylands. For many traders, the alternative to informal trade is not formal trade but rather no trade at all. Given the scale of informal trade in dryland regions, and considering the many barriers and costs associated with diverting goods to pass through formal border crossings, a better understanding of the unique challenges faced by traders is essential. Rather than criminalizing informal commerce, which merely serves to drive it underground, it would be advisable to provide traders with safer conditions. Delivering transparent and predictable trade rules and procedures, decreasing corruption at the border, together with instituting training and measures to improve access to information and to finance will address the key underlying causes of informality and provide a route for successful informal entrepreneurs (many of whom are women) into the formal economy.

## Note

1.  In their analysis of transport prices and costs throughout Africa, Teravaninthorn and
    Raballand (2008, p. 8) see the presence of cartels as central to high transport costs,
    but they argue that "deregulating the trucking industry in West and Central Africa
    is less a technical than a political and social issue. The main concern is that under a
    liberalized, competitive market, the demand could be served efficiently by a much
    smaller number of trucks."

## References

AfDB (African Development Bank). 2013. "African Economic Outlook 2013." African
    Development Bank, Tunis.

Aklilu, Y., P.D. Little, H. Mahmoud, and J. McPeak. 2013. "Market Access and Trade
    Issues Affecting the Drylands in the Horn of Africa." Brief prepared by a Technical
    Consortium hosted by CGIAR in partnership with the FAO Investment Centre.

Anderson K., M. Ivanic, and W. Martin. 2013. "Food Price Spikes, Price Insulation, and
    Poverty," Policy Research Working Paper 6535, World Bank, Washington, DC.

Badiane, O., S. Odjo, and S. Jemaneh, eds. 2014. "More Resilient Domestic Food Markets
    Through Regional Trade." In Ousmane Badiane, Tsitsi Makombe, and Godfrey
    Bahiigwa, editors, *Promoting Agricultural Trade to Enhance Resilience in Africa*.
    ReSAKSS 2013 Annual Trends and Outlook Report, 38–53. Washington, DC:
    International Food Policy Research Institute.

Badu, S.C. 2013. "Policy Process and Food Price Crisis: A Framework for Analysis and
    Lessons from Country Studies." WIDER Working Paper No. 2013/070. United Nations
    University World Institute for Development Economics Research (UNU-WIDER),
    Helsinki.

Brenton, P., A. Portugal-Perez, and J. Regolo. 2014. "Food Prices, Road Infrastructure,
    and Market Integration in Central and Eastern Africa." Policy Research Working
    Paper WPS 7003. World Bank, Washington, DC.

Brooks, K., S. Zoriya, and A. Gautam. 2012. "Employment in Agriculture: Jobs for
    Africa's Youth." In *Global Food Policy Report*, chapter 5. Washington, DC:
    International Food Policy Research Institute. http://www.ifpri.org/gfpr/2012/
    employment-agriculture

Brunelin, S., and A. Portugal-Perez. 2013. "Food Markets and Barriers to Regional
    Integration in West Africa." Unpublished document, Africa Region, World Bank,
    Washington, DC.

Ellis, F., and E. Manda. 2012 "Seasonal Food Crises and Policy Responses: A Narrative
    Account of Three Food Security Crises in Malawi." *World Development* 40(7):
    1407–17.

Giovannucci, D., P. Varangis, and D.F. Larson. 2000. "Warehouse Receipts: Facilitating
    Credit and Commodity Markets." Social Science Research Network. Available at
    SSRN: http://ssrn.com/abstract=952596.

Gisselquist, D., C.E. Pray, L. Nagarajan, and D.J. Spielman. 2013. "An Obstacle to Africa's Green Revolution: Too Few New Varieties." Available at SSRN: http://ssrn.com/abstract=2263042or http://dx.doi.org/10.2139/ssrn.2263042.

Gitonga, Z.M., H. De Groote, M. Kassie, and T. Tefera. 2013. "Impact of Metal Silos on Households' Maize Storage, Storage Losses and Food Security: An Application of a Propensity Score Matching." *Food Policy* 43: 44–55.

Haggblade, S. 2013. "Unscrambling Africa: Regional Requirements for Achieving Food Security." *Development Policy Review* 31(2): 149–76.

Harris, D., V. Chambers, and M. Foresti. 2011. "Final Report: The Political Economy of Regional Integration and Regionalism in West Africa: Scoping Study and Prioritisation." Overseas Development Institute (ODI), London.

Headey, D. 2010. "Rethinking the Global Food Crisis: The Role of Trade Shocks." *Food Policy* 36(2): 136–46.

Jayne, T. S., and D. Tschirley. 2009. "Food Price Spikes and Strategic Interactions Between the Public and Private Sectors: Market Failures Or Governance Failures?" Conference Proceedings, Institutions and Policies to Manage Global Market Risks and Price Spikes in Basic Food Commodities, 26–27. Trade and Markets Division, FAO Headquarters, Rome.

Jouanjean, M.A. 2013. "Targeting Infrastructure Development to Foster Agricultural Trade and Market Integration in Developing Countries: An Analytical Review." ODI, London.

Karapinar, B., and C. Haberli. 2010. *Food Crises and the WTO*. World Trade Forum. Cambridge, UK: Cambridge University Press.

Larson, D.F., J. Lampietti, C. Gouel, C. Cafiero, and J. Roberts. 2014. "Food Security and Storage in the Middle East and North Africa." *World Bank Economic Review* 28(1).

Magrini E., P. Montalbanob, S. Nenci, and L. Salvatici. 2013. "Agricultural Trade Distortions During Recent International Price Spikes: What Implications for Food Security?" European Trade Study Group (ETSG), Birmingham, U.K., 15th Annual Conference. University of Birmingham, September 12–14.

Martin W., and K. Anderson. 2012. "Export Restrictions and Price Insulation During Commodity Price Booms." *American Journal of Agricultural Economics* 94(2): 422–27.

Minot, N. 2012. "Food Price Volatility in Africa: Has It Really Increased?" IFPRI Discussion Paper 1239, International Food Policy Research Institute (IFPRI), Washington, DC.

Mitra, S., and T. Josling. 2009. "Agricultural Export Restrictions: Welfare Implications and Trade Disciplines." IPC Position Paper, Agricultural and Rural Development Policy Series. International Food Agricultural Trade Policy Council, Washington, DC.

Pannhausen, C., and B. Untied. 2010. "'Regional Agricultural Trade in West Africa: A Focus on the Sahel Region." GTZ Working Paper, Berlin.

Rutten M., L. Shutes, and G. Meijerink. 2011. "Sit Down At the Ball Game: How Trade Barriers Make the World Less Food Secure." *Food Policy* 38: 1–10.

Slater, R., and J. Dana. 2006. "Tackling Vulnerability to Hunger in Malawi through Market-based Options Contracts." *Humanitarian Exchange Magazine* 33. http://www.odihpn.org/report.asp?id=2790.

Tefera, T. 2012. "Post-Harvest Losses in African Maize in the Face of Increasing Food Shortage." *Food Security* 4: 267–77.

Teravaninthorn, S., and G. Raballand. 2008. "Transport Prices and Costs in Africa: A Review of the Main International Corridors." Africa Infrastructure Country Diagnostic (AICD) Working Paper 14, Sustainable Development Department, Africa Region. World Bank, Washington, DC.

USAID (U.S. Agency for International Development). 2011. "Agribusiness and Trade Promotion and Expanded Agribusiness and Trade Promotion (USAID EATP). Annual Progress Report October 2010–September 2011." USAID, Washington, DC.

Wiggins, S., and S. Keats. 2013. "Looking Back, Peering Forward: What Has Been Learned From The Food Price Spike of 2007–2008?" ODI Briefing 81, Overseas Development Institute (ODI), London.

## Chapter 10

# Social Protection: Building Resilience among the Poor and Protecting the Most Vulnerable

*Carlo del Ninno, Sarah Coll-Black, Pierre Fallavier*

## Current situation

Social protection programs have been proliferating and expanding in size in recent years, reflecting their increased acceptance among policy makers as a key component of any national poverty reduction strategy (for a definition of social protection programs, see box 10.1). Yet despite their growing popularity worldwide, social protection programs in dryland regions of Africa remain underfunded compared to those in other developing regions, and as a result their coverage remains limited. In the Horn of Africa and the Sahel, as in other parts of Africa, most social protection programs are small, fragmented, and largely donor-driven. Still, countries such as Ethiopia, Kenya, and more recently Uganda have scaled up their investments in social protection with encouraging results, providing a model for how other countries can progressively expand coverage to poor, vulnerable populations.

### Coverage of social protection programs

In East and West Africa, as in other parts of the world, the oldest form of social protection programs are national social security schemes. These schemes typically provide pensions for civil servants and those employed in the formal private sector. Despite their long history, national social security schemes typically cover only a fraction of the population and generally do not provide effective protection from poverty in old age or following adverse lifecycle events. In many countries of the region, social security schemes are underresourced, both in terms of staff and funding.[1] Even so, they can consume significant resources. In Kenya, for instance, expenditure on civil service pensions in 2010 represented about 1 percent of GDP and 88 percent of total government spending on social protection. In Uganda projections show that over time government

BOX 10.1

## Defining social protection

Social protection systems, programs, and policies help disadvantaged people recover from shocks and take advantage of opportunities to improve their livelihoods. They do this by providing basic income support to the poor, to help them cope with the impacts of adverse events and build the resources needed for a more prosperous and resilient future. Social protection can be provided in different ways through various instruments as follows:

**Pension systems** provide income during old age. Pension schemes may be contributory or noncontributory (the latter are known as "social pensions").

**Insurance** is designed to protect the well-being of individuals, households, and businesses in the face of adverse events, particularly those affecting primary livelihood activities, such as livestock-keeping and crop farming.

**Labor programs and policies** promote productive employment in the formal and informal sectors. Common interventions include initiatives to enhance the skills of the workforce and to support entrepreneurship and self-employment, particularly among youth.

**Safety nets** are noncontributory transfer programs targeted to the poorest and most vulnerable. They include such things as cash transfers, public works, and in-kind support (e.g., fee waivers and school feeding).

*Source:* World Bank 2012.

expenditure on the public service pension scheme is likely to increase more than threefold, to 1.1 percent of GDP.[2]

Safety net programs that aim directly to reduce poverty and vulnerability began to appear in the Horn of Africa and the Sahel around 2005. Unlike social security schemes, these initiatives are designed to respond to current high levels of chronic poverty and vulnerability, rather than to ensure future income streams against the loss of employment in old age or as a result of adverse life-cycle events. In many countries safety nets were introduced as an alternative to annual distribution of emergency food aid. More recently there has been a proliferation of safety nets, many put in place to provide short-term responses to acute humanitarian needs. The use of food transfers remains common in many countries. In South Sudan, for example, 98 percent of beneficiaries receive safety net support in the form of food.

In Ethiopia, Kenya, Uganda, and most recently Djibouti, there has been a strong push to establish national safety net programs. In the Sahel, not until 2010 were safety net initiatives seen as an approach that could be used on a large scale. Despite the recent trend toward setting up national programs, spending on safety net programs in the Horn of Africa and the Sahel remains generally low, even in comparison to other countries in Africa (figure 10.1). Within the general trend, there is enormous variation across countries, reflecting different scales of coverage, varying levels of payments to beneficiaries, different payment modalities (e.g., cash versus food), and highly variable administrative costs. In all of these countries, a significant proportion of funding for safety net programs is provided by donor agencies.

Beyond cash transfer programs, few social protection interventions have been used in the drylands of Africa. A growing body of evidence suggests that insurance and labor programs can assist households to better mitigate the impact of shocks and diversify their livelihoods, yet such programs have rarely been introduced in drylands, and those that exist tend to be small pilot initiatives.

## Capacity of national programs to respond to vulnerability in the drylands

As of 2013 only three safety net programs in the Horn of Africa could be characterized as being national in scale, with the Ethiopia Productive Safety Net Program (PSNP) by far the largest (World Bank 2010). Even in countries with relatively well-established safety net programs, coverage is low in relation to the size of the population needing support. In Ethiopia, PSNP reaches less than 7 percent of the population (roughly 24 percent of the poor), while in Kenya cash transfer programs provided support for up to 15 percent of the absolute poor population in 2014.[3] Since safety net programs in the Horn of Africa and the Sahel have historically been used to target chronic poverty, the coverage of safety nets is often higher in dryland areas than in other areas. For example in Kenya, because the government has tried to prioritize poorer areas when extending safety net support, coverage rates among absolute poor

**Figure 10.1** Government and donor spending on social safety nets as a share of GDP, selected countries (%)

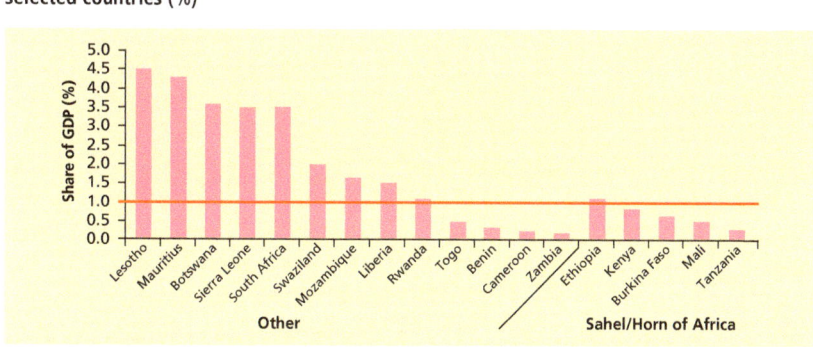

*Source:* Monchuk 2014.

*Note:* The reference line (1 percent of GDP) indicates the consensus value in the social protection literature on the resources governments should be willing to spend in social safety nets.

households in the four arid- and semi-arid counties in northern Kenya exceed 40 percent.[4]

In the Sahelian countries the coverage of safety net programs is more limited. Although some safety net programs reach a large percentage of the population, recipients may not be those most in need of support, and as a result, resources are inefficiently used. An example can be seen in programs involving large-scale distribution of free food or subsidized food. In Burkina Faso, for example, although the resources invested in food subsidies theoretically are sufficient to assist more than 3.9 million people (60 percent of the poor), surveys suggest that the number of poor who actually receive subsidized food is much smaller. In Senegal the effective coverage may be even lower: 80 percent of the 4 million people receiving some type of safety net assistance obtain it through the national food aid system, which distributes free food without proof of need.[5]

In response to the inefficiencies of earlier safety net systems, more recently a number of countries have begun experimenting with new models. In 2011 Niger began to provide chronically poor households with regular cash transfers over an 18–24 month period, with the goal of helping them meet basic consumption needs while gradually building their human capital. Over time, similar programs were introduced in other Sahelian countries. The model is simple—cash is provided along with accompanying measures, such as education, to raise nutritional awareness among mothers, or training to instill employable skills among working-age youth and adults. The programs are designed to be flexible, so that the amounts of the transfers and the types of accompanying measures can be adapted to local needs and so that coverage can be scaled up in times of crisis. While this new generation of safety net programs shows signs of promise, most of the programs are still at early stages of implementation and are not ready to be scaled up rapidly in response to a crisis. The experience in Niger was generally positive, although an important lesson that has emerged is that a one-size-fits-all approach is not always effective, as permanent programs and emergency responses need to be adapted to the diversity of livelihood systems found throughout the country (map 10.1).

Several countries in East Africa have made efforts to tailor safety net programs to meet local needs. In Kenya, for example, the Hunger Safety Net Program (HSNP) was designed specifically to respond to the vulnerabilities of people living in the arid and semi-arid areas of the northern part of the country. HSNP uses cell phone–based technology to support a mobile payment system that is adaptable to pastoral livelihoods. Under PSNP In Ethiopia efforts have been made to tailor the design and delivery of assistance to the pastoral regions of Afar and Somali. These efforts were launched within the parameters of an existing program, however, and because certain features of the preexisting program proved inflexible, the results were mixed. Despite the variable results, PSNP provides a rare example of a safety net program that has attempted to tailor the design and delivery of public works to pastoral livelihoods (Lind and Kohnstamm 2014; World Bank 2010).

**Map 10.1** Diversity of rural livelihoods in Niger

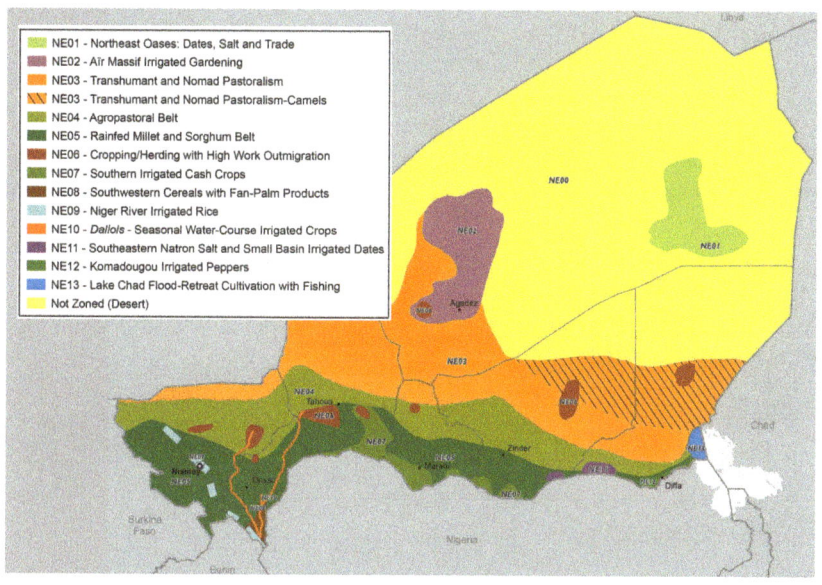

- NE01 - Northeast Oases: Dates, Salt and Trade
- NE02 - Aïr Massif Irrigated Gardening
- NE03 - Transhumant and Nomad Pastoralism
- NE03 - Transhumant and Nomad Pastoralism-Camels
- NE04 - Agropastoral Belt
- NE05 - Rainfed Millet and Sorghum Belt
- NE06 - Cropping/Herding with High Work Outmigration
- NE07 - Southern Irrigated Cash Crops
- NE08 - Southwestern Cereals with Fan-Palm Products
- NE09 - Niger River Irrigated Rice
- NE10 - *Dallols* - Seasonal Water-Course Irrigated Crops
- NE11 - Southeastern Natron Salt and Small Basin Irrigated Dates
- NE12 - Komadougou Irrigated Peppers
- NE13 - Lake Chad Flood-Retreat Cultivation with Fishing
- Not Zoned (Desert)

*Source:* HEA Sahel http://www.hea-sahel.org, retrieved January 2014.

Among African countries, only Ethiopia has established the capacity to expand the coverage of its safety net program rapidly in response to shocks. This capacity is critically needed in dryland zones, where large numbers of poor people are chronically exposed to droughts that can suddenly undermine their livelihood strategies. In Ethiopia rapid scalability of PSNP is ensured through contingency funds that are held at district (*woreda*) and regional levels. These contingency funds can be used by local officials to respond to transitory food insecurity, including food insecurity arising from drought. Since 2008 the contingency funds have been complemented with a risk financing mechanism, which allows the federal government to trigger the release of additional resources to increase the value or frequency of transfers to existing beneficiaries and to provide support to additional people who have been negatively affected by drought. The scalability feature of PSNP was designed to provide a first line of response to drought, complementing the existing humanitarian appeal mechanism, which will continue to be used to respond to needs in areas outside PSNP districts or in cases where needs within PSNP districts exceed available resources. During the 2011 Horn of Africa crisis, the administrative and logistical infrastructure of PSNP proved capable of scaling up the coverage of the program very rapidly, thereby strengthening the capacity of hundreds of thousands of vulnerable households to withstand a series of unexpected shocks.

## Continued reliance on humanitarian response

Over the next two decades and beyond, large numbers of vulnerable people will continue to be exposed to droughts in the drylands of Africa (table 10.1). Given the current limitations in social protection programming, humanitarian assistance is likely to remain a major form of support to households in these areas.

Because social protection programs in dryland countries are generally very small, and because few of the existing programs have the capacity to scale up in response to shocks, during times of crisis most governments continue to rely on humanitarian appeals (figure 10.2). The value of humanitarian assistance to the Sahel increased from US$37 million in 2000 to US$630 million in 2010. In 2014

**Table 10.1** Projected evolution of vulnerability 2010–30 among agriculture-dependent population in drylands, for different GDP growth scenarios

| Population living under $1.25 per person per day (in million people) | Baseline 2010 | 2030 low GDP growth | 2030 average GDP growth | 2030 high GDP growth |
|---|---|---|---|---|
| **East Africa** | **25.18** | **42.39** | **31.81** | **22.85** |
| Ethiopia | 9.96 | 18.73 | 12.04 | 6.80 |
| Kenya | 3.72 | 5.19 | 4.50 | 4.13 |
| Uganda | 1.79 | 2.70 | 2.00 | 1.27 |
| Tanzania | 9.71 | 15.78 | 13.27 | 10.65 |
| **West Africa** | **42.22** | **86.89** | **69.53** | **55.42** |
| Benin | 1.07 | 1.03 | 0.80 | 0.49 |
| Burkina Faso | 5.53 | 6.61 | 5.55 | 4.46 |
| Chad | 2.80 | 8.03 | 3.99 | 3.03 |
| Côte d'Ivoire | 0.82 | 1.26 | 1.25 | 1.05 |
| Gambia, The | 0.37 | 0.77 | 0.55 | 0.63 |
| Ghana | 0.84 | 0.99 | 0.46 | 0.08 |
| Guinea | 0.17 | 0.26 | 0.22 | 0.23 |
| Guinea-Bissau | 0.02 | 0.03 | 0.04 | 0.03 |
| Mali | 3.57 | 6.10 | 5.48 | 4.85 |
| Mauritania | 0.45 | 0.77 | 0.60 | 0.43 |
| Niger | 4.41 | 16.96 | 15.18 | 13.65 |
| Nigeria | 19.12 | 37.98 | 29.90 | 21.56 |
| Senegal | 1.95 | 3.90 | 3.51 | 3.12 |
| Togo | 1.09 | 2.20 | 2.00 | 1.81 |
| **Grand Total** | **67.40** | **129.27** | **101.34** | **78.27** |

*Source:* Calculation based on the approach discussed in the Appendix.

*Note:* Countries without drylands were excluded: Djibouti, Eritrea, Liberia, Sierra Leone, Somalia, South Sudan, and Sudan.

**Figure 10.2** Humanitarian aid received, selected countries, Horn of Africa and Sahel, 2000–11 (US$ million)

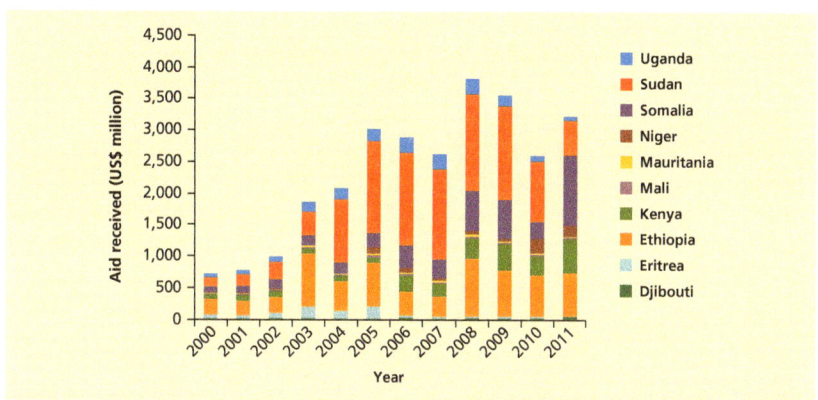

*Source:* Data from Global Humanitarian Assistance (http://www.globalhumanitarianassistance.org/data-guides/datastore).

the value of humanitarian assistance totaled US$878 million (45 percent of the estimated needs of US$1.95 billion), and for 2015 the UN Office for the Coordination of Humanitarian Affairs (OCHA) evaluated the humanitarian needs at US$1.96 billion. Of this, about 50 percent would be for food security and nutrition, and about 20 percent for support to displaced people and refugees (UNOCHA 2015).

Humanitarian assistance in the drylands typically involves the provision of food, cash, and other in-kind resources and services to help affected households cope with the immediate effects of drought. Delivery mechanisms for humanitarian aid often consist of food distribution schemes, cash transfer programs, feeding programs, purchase of livestock, and provision of health services and water and sanitation services. Humanitarian assistance is an appropriate short-term response to emergencies, but in many countries it is provided year after year in the same areas and to the same recipients, suggesting it is being used as a long-term instrument to address chronic poverty. This use of humanitarian aid is inappropriate, because the delivery costs tend to be extremely high. Food aid, for example, is usually procured internationally and transported across long distances, making it very expensive. In Ethiopia prior to 2005 when PSNP was introduced, food distribution programs had become the annual response to chronic food insecurity, costing on average US$265 million per year. In Kenya from 2005 to 2010 spending on food aid accounted for 53.2 percent of all government spending on safety nets. Given the high cost of delivering food aid, it is estimated that every dollar spent on food aid could have generated twice as many benefits to recipients had it been provided in the form of a permanent cash transfer program.

In addition to being expensive, protracted use of humanitarian assistance is often ineffective. While emergency distribution of food can save lives, the implementation challenges are considerable. The food often arrives late, and the amount delivered is generally less than what is required. Additionally, given the emergency nature of the support, it is often difficult to target the poorest and most vulnerable households; the authorities tend to focus mainly on getting the resources to communities that have been especially hard hit, but the allocation of resources to households within these communities is often done in an ad hoc fashion, or the resources are made available to all households regardless of need. Finally, because humanitarian aid resources become available only after a shock has occurred and donors have had time to respond to appeals, the timing and amount of transfers received by the affected households tend to be inadequate to meet all of their needs.

## Opportunities

Social protection programs, when correctly designed and effectively implemented, can reduce vulnerability in the drylands by reducing the sensitivity to shocks of vulnerable households and improving the capacity of these households to cope after a shock has occurred. When designing interventions, however, it is important to distinguish between these two objectives (reducing sensitivity and improving coping capacity) and to consider the characteristics— including the financing requirements—of the different types of interventions that can be used to achieve each objective.

### Reducing sensitivity

Social protection programs can reduce sensitivity to shocks by enabling poor and vulnerable households to invest in human capital, build assets, and diversify their livelihood strategies. The social protection programs that perform this function are those that target the chronic poor and provide continuous assistance over a sustained period. Sustained, predictable support provides the certainty households need to be able to take risks that can lead to higher returns on investments and enhanced income streams. Predictable, multi-annual social protection support to households has also been shown to stimulate investments in human capital and assets that can, over the longer term, lift households out of poverty. While the assistance is provided over multiple years, the expectation is that for individual households it is finite, in the sense that it will be suspended once the household has built an asset base and diversified its livelihood strategy, because at that point the household will be resilient and will no longer require support. These objectives are more effectively achieved when social protection support is combined with investments in human capital and livelihoods, and when it is integrated with other development programs, such as those that are being proposed for dryland areas.[6]

**Cash transfers.** One type of social protection program that allows households to invest in human capital, build assets, and diversify their livelihood strategy involves cash transfers. Cash transfers may be unconditional or conditional. Unconditional cash transfers provide greater flexibility for recipients to use the money to address their own priorities, but they bring the risk that the resources may be used for immediate consumption instead of being invested in ways that would allow the recipients to improve their livelihoods in future years. Conditional cash transfers are designed to overcome this problem by requiring recipients to engage in activities that are likely to generate benefits over the longer term. Increasingly, the delivery of support is complemented with other services, such as those that promote nutrition or provide skills training. This approach is becoming particularly common in the Sahel. When properly designed, these programs can support more productive and potentially more diversified livelihoods, and they can help people participate in the growth process by taking advantage of the types of investments in livestock production, agriculture, and irrigation described elsewhere in this book.

A large and growing body of evidence shows that cash transfer programs work, including in dryland areas. In the arid and semi-arid zones of northern Kenya, households receiving regular cash transfer support from the HSNP withstood a severe drought in 2011 without any increase in poverty levels, whereas among those not receiving cash transfer support, 5.3 percent of households fell into the bottom income decile following the drought. In Ethiopia the average period during which households participating in PSNP reported being food-secure increased from 8.4 months in 2006 to 10.1 months in 2012. While it is not possible to disaggregate these findings by aridity zone, data from regions in Ethiopia that are predominantly classified as drylands show results that are similar to those recorded in more humid regions (Hoddinott and Lind 2013).

**Public works.** A second type of social protection program that can help households reduce their sensitivity to shocks is public works. In addition to delivering immediate assistance to participating households by paying wages, public works can put in place productive infrastructure that can improve permanently the livelihood strategies of recipients. Public works programs are particularly common throughout the Horn of Africa. More than a decade of experience with public works programs in Ethiopia has demonstrated how watershed development schemes have the potential to transform the natural environment and enhance the resilience of communities and households, especially when they are designed using community-based planning approaches and implemented over multiple years. Through public works initiatives, PSNP has constructed 600,000 kilometers (km) of soil and stone bunds that enhance water retention and reduce soil erosion. Public works initiatives supported under PSNP have also been used to protect 644,000 hectares of land in area enclosures, leading to improved soil fertility and increased carbon sequestration. Within these enclosures, groundwater levels are rising, springs last longer into the dry season, and wood and herbaceous vegetation has increased. These results are having a direct impact on rural livelihoods (World Bank 2014).

**Insurance programs.** A third type of social protection instrument that can reduce sensitivity to shocks is a program that facilitates access to insurance products that lower the risk associated with traditional livelihood strategies, such as farming and livestock-keeping. Typically these products are designed to provide protection against extreme weather events, including drought, by linking payouts to weather-based indexes. While these have been tested only on a limited scale through pilot schemes, experience suggests that well-designed weather indexed insurance programs can be effective in protecting rural households from shocks. In Kenya, for example, when drought triggered payouts by the Index-Based Livestock Insurance, the frequency with which households protected by the scheme engaged in negative coping strategies (such as selling livestock or reducing the number of meals eaten each day) fell by 33 percent, and the frequency with which they engaged in distress sales of livestock fell by 50 percent. A 33 percent drop in food aid reliance was also observed. In Ethiopia evaluations of households insured through the Rural Resilience Initiative concluded that compared with non-participants, farmers who bought insurance planted more seeds, used more compost, adopted modern varieties at higher rates, used less family labor and more hired labor, diversified their income sources, and experienced smaller losses of livestock (Hoddinott and Lind 2013). If the experience gained through these pilot schemes can be harnessed to build effective large-scale insurance programs, the coping capacity of households living in drylands could be further strengthened. Over time as they become confident that insurance products can provide effective protection against the negative effects of shocks, households will be encouraged to invest in more productive livelihood strategies that will reduce their chances of falling into poverty.

### Improving coping capacity
In addition to reducing sensitivity to shocks, social protection programs can improve coping capacity and help households recover after a shock has hit by providing immediate assistance, usually in the form of food or money. Unlike other types of social protection programs that target the chronic poor and provide continuous assistance over a sustained period, this second type of program—often referred to as a "temporary" safety net—is designed to provide short-term assistance to help affected households cope with the effects of a specific shock. Unlike other types of programs that are designed to encourage

households to invest in human capital, build assets, and diversity their livelihoods, this type of program allows immediate needs to be met by providing consumption support, thereby allowing households to avoid the use of short-run, negative coping strategies that will undermine their livelihoods over the longer term, such as selling livestock or pulling their children out of school.[7] It is important to note that this type of program is not expected to have a permanent effect on the poverty status of beneficiary households, although these households may avoid falling deeper into poverty. Households that receive benefits through this type of program will be made resilient in the year in which they receive the benefits, but they will not necessarily be resilient in subsequent years, after the flow of benefits has stopped.

Because this second type of program is designed to improve coping capacity by taking action when a shock is imminent or after a shock has hit, it is critically important that whatever instruments are used be part of the permanent system and that they be rapidly scalable. In addition, it is important that scalable safety net programs be linked clearly to humanitarian support, so that humanitarian support can be mobilized quickly when the capacities of scalable safety net programs are exceeded.

## National safety set programs

The core of any successful safety net system is the ability to scale up coverage rapidly and efficiently. Currently in Africa safety nets are at different stages of development (table 10.2). A number of countries in the Horn of Africa, including Ethiopia, Kenya, and Uganda, have made the most progress in putting in place national safety net programs. While the rationales for these programs and their features differ, each country has established a government-led safety net program that is national in scope. These initiatives can serve as examples to the many Sahelian countries that have yet to introduce safety net programs, as well as Somalia, Sudan, and South Sudan, whose investments in safety net programs have been modest.

Since the incidence, severity, and impacts of many shocks cannot be predicted, scalability is of paramount importance in the design of safety nets. To be effective, a national safety net program must be capable of rapidly expanding the provision of transfers to people who have been (or are about to be) negatively affected by a shock. The best scalable safety nets are able to respond quickly to an imminent or emerging crisis on the basis of information generated through early warning systems and seasonal assessments.

Scaling up of existing safety net programs allows for a much faster response to drought and other emergencies than is possible using the traditional humanitarian appeal process. Additionally, transfer systems that are already in place can have a greater impact in terms of consumption smoothing and livelihood protection per dollar spent than expensive ad hoc programs. Investing in early

**Table 10.2** Country typology based on crisis preparedness and SSN capacity

|  | Strong measures to improve SSNs during a crisis | Moderate measures to improve SSNs during a crisis | Limited or no measures to improve SSNs during a crisis |
|---|---|---|---|
| **Tier I**<br>**No SSNs in place** |  | Comoros | Central African Republic, **Chad,** Congo, Côte d'Ivoire, Equatorial Guinea, **Eritrea,** The Gambia, Guinea, **Mauritania, Somalia, South Sudan, Sudan** |
| **Tier II**<br>**Weak capacity in SSNs** | **Niger,** Tanzania, Zimbabwe | Ghana, Liberia, Malawi, Mozambique, Sierra Leone, Togo, **Uganda** | Angola, Benin, Burkina Faso, Burundi, Cameroon, the Democratic Republic of Congo, Gabon, Guinea Bissau, Madagascar, **Mali, Nigeria,** São Tomé and Príncipe, **Senegal,** Swaziland, Zambia |
| **Tier III**<br>**Increasing capacity in SSNs** | **Ethiopia, Kenya,** Rwanda | Cape Verde, Lesotho, Mauritius |  |
| **Tier IV**<br>**High capacity in SSNs** |  | Botswana, Namibia, South Africa |  |

*Source:* Adapted from Monchuk 2014.

*Note:* Countries in bold type are located in dryland regions of the Sahel or the Horn of Africa. SSN = social safety net.

warning systems is central to this approach, to ensure that there is a reliable and transparent stream of information as the basis for triggering any response.

Recent innovations in delivery mechanisms, particularly the use of information and communications technology (ICT), offer opportunities to reach remote populations, which is of particular interest to dryland regions. In northern Kenya investments in solar panels and smart card technology have enabled the HSNP to create a payment system that is responsive to the mobile lifestyles of pastoral populations. In Somalia mobile phone technology has played an important role in the Shaqodoon initiative, which uses interactive Somali-language audio programs on financial literacy and entrepreneurship to link youth to employment opportunities vial mobile phones and the Internet (Lind and Kohnstamm 2014).

National safety net programs are often thought to be expensive, but in considering the cost it is important not to lose sight of the cost of alternative interventions used to achieve the same objectives. Extending coverage of an existing social safety net program is usually much more cost-effective than relying on ad hoc humanitarian responses in times of crisis (table 10.3). For example in Kenya, reorienting existing spending on general food distribution or food aid (estimated to cost US$61 million per year) would double the current levels of

**Table 10.3** Cost of SSN support to poor households compared to humanitarian responses, selected countries, 2010–13 (US$)

| | Annual cost of regular safety net support to bottom quintile (US$) | | | | | Avg. cost of humanitarian response, 2010–13 |
|---|---|---|---|---|---|---|
| | Hyper-arid | Arid | Semi-arid | Dry subhumid | Total | |
| Burkina Faso | 0 | 1,371,749 | 88,833,727 | 11,782,273 | 101,987,750 | 48,555,902 |
| Chad | 781,398 | 17,128,141 | 48,718,180 | 17,214,163 | 83,841,882 | 298,148,319 |
| Mali | 210,643 | 14,643,841 | 69,557,074 | 16,788,531 | 101,200,089 | 77,423,890 |
| Mauritania | 3,107,358 | 15,568,742 | 825,661 | 0 | 19,501,761 | 34,784,819 |
| Niger | 1,681,344 | 52,017,414 | 48,277,168 | 0 | 101,975,926 | 218,221,834 |
| Senegal | 0 | 9,016,207 | 66,455,931 | 7,781,703 | 83,253,841 | 7,357,294 |
| Total | 5,780,743 | 109,746,094 | 322,667,740 | 53,566,670 | 491,761,248 | 684,492,057 |

*Source:* Calculations based on World Bank data.

*Note:* Number of poor households calculated based on the national poverty line of each country. Annual cost of safety net support estimated to be US$300 per household. SSN = social safety net.

financing available for cash transfers and make possible high rates of coverage of poor and vulnerable households. In Ethiopia since PSNP was launched in 2005 the government has received US$623.6 million per year on average for humanitarian responses, an amount that if allocated to PSNP could extend regular support to a significant proportion of the population living below the poverty line. In Niger providing regular cash transfer support to the poorest 20 percent of the population would cost US$83 million per year, compared with an average of US$218 million per year spent on humanitarian responses in the period 2010–13. The intuition emerging from these experiences is confirmed in a recent comparative study by Venton et al. (2012), who found that building resilience and taking early action is far more cost-effective than relying on late humanitarian responses.

National safety net programs may be cost-effective relative to humanitarian responses, but they can still require a significant commitment of resources—with the size of the commitment depending on the scope of coverage and the level of support provided. In a world of unlimited resources and perfect targeting, national safety net programs theoretically could be used to make all drought-affected households in drylands resilient by providing them with cash transfers in the amounts needed to bring every household up to the poverty line. Alternatively the level of support provided to each household could be scaled back, with the objective of reaching larger numbers of people. Figure 10.3 shows the estimated cost in 2030 of providing safety net support to drought-affected people at two levels of support in selected dryland countries, expressed as a percentage of GDP. The cost of bringing to the poverty line all drought-affected households ranges from less than 0.5 percent of GDP in countries with

**Figure 10.3** Cost of ensuring resilience through safety net support as share of GDP, selected countries, 2030 (%)

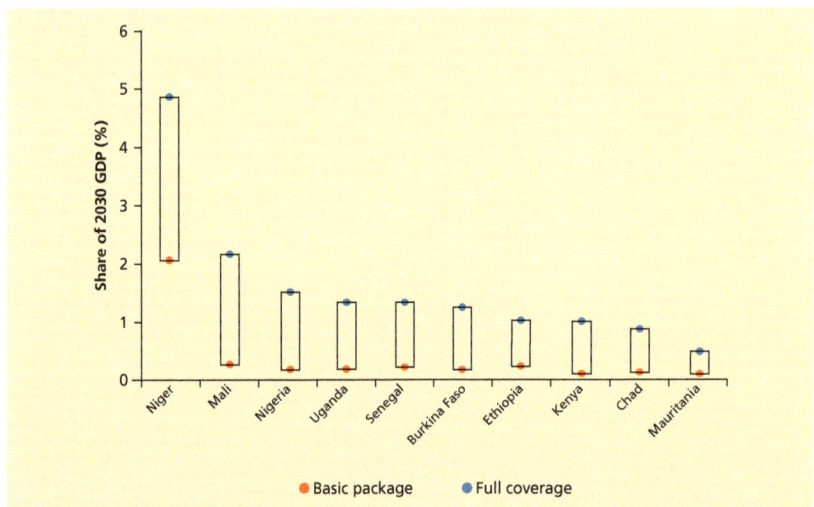

*Source:* Calculation based on the approach discussed in the Appendix.

relatively high GDP per capita (e.g., Mauritania) to almost 5 percent of GDP in in countries with relatively low GDP per capita and extensive dryland populations (e.g., Niger). The cost of providing to all drought-affected households a minimum assistance package worth US$50/person/year (or US$600/household of six people/year[8]) is more modest, ranging from around 0.1 percent of GDP to around 2.1 percent of GDP. A minimum assistance package of this amount, which is close to the historical average in the Horn of Africa and the Sahel, will not have any measurable effect in terms of building resilience over the longer term, but experience shows that it can at least help drought-affected populations smooth their income and avoid engaging in negative coping mechanisms until the following year.

Policy makers do not live in a world of unlimited resources, and in many dryland countries, investments in safety net programs, even at the lower of these two levels, are not possible. While there is no simple golden rule concerning the amount of funding that should be committed to social protection programs, the need for which can vary tremendously from country to country, many in the development community believe a reasonable reference level of support to social safety nets system should be 1 percent of GDP per year.

The umbrella model was used in the present study to estimate the potential extent of safety net coverage in the Horn of Africa and the Sahel in 2030 if every country were to invest 1 percent of GDP annually in social safety nets programs.

As shown in figure 10.4, with the exception of Mauritania and Chad, in many dryland countries earmarking 1 percent of GDP to social safety nets would be sufficient to provide full protection to people affected by drought in an average year.

In assessing these findings, it is important to keep in mind that the households benefiting from social protection programs fall into two very different groups. The first group consists of chronically poor households that receive assistance through social protection programs designed to help them meet their basic consumption needs. Once these basic needs have been met, they are able to invest in other things, such as health and education, which allows them to acquire skills and build the assets needed to emerge from poverty over the longer term. These households also are better able to cope in the year in which they receive the assistance (that is, they will be resilient in that year), but crucially their vulnerability status is likely to change permanently after they have participated in the program for some time. The second group consists of transiently poor and chronically poor households that receive assistance through safety net–type programs designed to help them recover from shocks in the short run. These households are better able to cope in the year in which they receive the assistance (that is, they will be resilient in that year), but their vulnerability

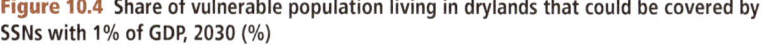

**Figure 10.4** Share of vulnerable population living in drylands that could be covered by SSNs with 1% of GDP, 2030 (%)

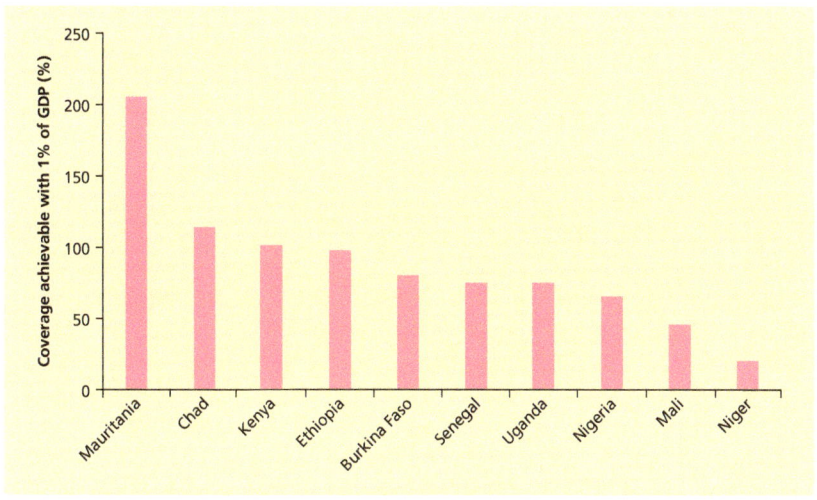

*Source:* Calculation based on the approach discussed in the Appendix.

*Note:* Drought-affected population includes people living in drylands, dependent on agriculture, and whose income would be below the poverty line following a drought. GDP figures are prorated based on the share of people living in drylands. SSN = social safety net.

status will not change for the positive in subsequent years, although it would also not become worse as a result of shocks.

Resources invested in social protection will have to be allocated between safety net programs designed to help chronically poor households meet their consumption needs and develop the skills and build the assets needed to emerge from poverty over the longer term, and those designed to help transiently poor households recover from shocks in the short run. From a development perspective, the first type of program is clearly preferable, but when a shock has occurred and people are suffering, political and humanitarian considerations will almost always demand that the second type be funded. The challenge for policy makers is to strike an appropriate balance between the two, a task made especially difficult by the fact that financing needs for safety nets are inherently unpredictable. However, the emerging experience with scalable safety nets suggests that investments made in permanent systems reduce the costs associated with delivery support to households negatively affected by drought.

## Challenges

Social protection systems can be efficient and cost-effective instruments for responding to crises in the drylands of Africa, but they must be well designed. Differences in the geographical distribution of the population, the nature of predominant livelihood strategies, and the depth of poverty call for different design and delivery mechanisms of social protection programs. A recurring question is whether vulnerable populations living in drylands will be adequately served by a single model applied uniformly across the entire country, or whether specialized policies and programs will be needed that are tailored to their special needs.

Challenges arise as well in determining trade-offs between programs that respond to the needs of the chronically poor, as opposed to programs that provide temporary support during periods of crisis to both chronic and transient poor. These tradeoffs are especially acute in the drylands, which are home to many chronically poor but also to many transient poor who fall in and out of poverty as a result of their exposure to frequent shocks. For example, in Ethiopia some evidence suggests that pastoral populations are generally better off than agricultural populations, but pastoral livelihoods are far more exposed to shocks and sensitive to shocks, so pastoral populations are far more likely to require periodic assistance. In cases such as this, policy makers face the difficult decision of how best to allocate resources between programs that respond to the needs of the chronically poor and programs that respond to the needs of the transiently poor (currently in PSNP, 80 percent of the resources go to the former).

Lack of government capacity can be a real constraint to extending the coverage of existing social protection programs. Capacity limitations are particularly acute in remote parts of the drylands, where the presence of government agencies is often limited. If social protection programs are to succeed in dryland regions, concerted efforts will have to be made to build implementation capacity, starting with the posting of qualified staff to decentralized locations. Effective delivery of insurance products will especially depend on the existence of well-functioning information systems and far-reaching financial networks, since insurance schemes require high-quality, reliable data, as well as decentralized systems for collecting and making cash payments.

## Key messages

**Social protection programs will be a key component of strategies to increase resilience and reduce vulnerability in the drylands.** If present trends continue, by 2030 dryland regions of East and West Africa will be home to an estimated 429 million people, up to 24 percent of whom will be living in chronic poverty. Many others will depend on livelihood strategies that are sensitive to the shocks that will hit the region with increasing frequency and severity, making them vulnerable to falling into transient poverty. Social protection programs thus will be needed in the drylands to provide support to those unable to meet their basic needs. Some of these people will require long-term support, while others will require periodic short-term support because of income losses due to shocks (for example, crop failure following a drought) or as a result of lifecycle changes (for example, loss of a breadwinner).

**Safety net programs can increase resilience in the short term by improving coping capacity of vulnerable households.** Rapidly scalable safety nets that provide cash, food, or other resources to shock-affected households can allow them to recover from unexpected shocks. Scaling up an existing safety net program can be far less expensive than relying on appeals for humanitarian assistance to meet urgent needs. Despite the fact that safety nets are a more effective response to poverty and vulnerability than emergency assistance, funding of safety nets is low, and flows of humanitarian resources to countries in the Horn and the Sahel remain high.

**Social protection programs can increase resilience over the longer term by reducing sensitivity to shocks of vulnerable households, especially if combined with other development programs.** Safety net programs must be complemented by other types of social protection programs that enable chronically poor households to build their productive assets and expand their income-earning opportunities. Providing predictable support to chronically poor households and enabling them to invest in productive assets and access basic social services can effectively reduce their sensitivity to future shocks and help them participate in the growth process and take advantage of investments made in improving existing livelihood strategies. Households covered by well-functioning social protection programs are less likely to resort to negative coping strategies, such as pulling their children out of school and selling productive assets.

**Safety net programs must be able to scale up in response to shocks.** The dynamic nature of vulnerability in dryland areas draws attention to the need for safety net programs to be able to scale up in the face of shocks and then to scale back down when these pass. In dryland areas such instruments may be even more important than in non-dryland areas, given the levels of vulnerability and exposure to shocks. Emergency support should be provided on an occasional basis whenever a set of predefined triggers are met and in a manner that complements, rather than replaces, the support extended through scalable safety nets. Effective early warning and monitoring systems are needed to alert policy makers and guide the response.

**Social protection programs must be tailored to address the unique circumstances of dryland populations.** The needs of poor households living in drylands often differ from those of poor households living in more favorable environments or in urban areas. For this reason, one-size-fits-all programs implemented at the national level often fail to adequately address the needs of dryland populations. Interventions designed to strengthen the livelihood strategies of dryland populations and build their resilience will not be effective if they fail to account for their specific needs. Program delivery mechanisms similarly need to respond to the specific needs of dryland populations, for example, by accommodating the mobility of pastoral populations.

**Capacity constraints will need to be overcome to ensure effective implementation of social protection programs in the drylands.** Effective implementation of social protection programs in the drylands is made difficult by the limited presence of public agencies and the lack of infrastructure. Incentives are needed to attract and retain qualified staff in hardship posts. Investments in transportation systems and information technology are needed to improve mobility and reduce transactions costs associated with implementing social protection programs in remote dryland areas.

**Investing in scalable safety net programs is extremely cost-effective over the longer term.** While it saves lives in the short run, humanitarian assistance generally does little to build resilience and help cushion the impacts of future shocks. Policy makers and development partners must find ways to redirect resources away from short-term emergency responses, including possibly inefficient humanitarian assistance, to build scalable safety net programs that will build the resilience of vulnerable populations and reduce future needs for emergency responses.

## Notes

1. In assessing the effectiveness of social security schemes in the Sahel, one fact stands out: In all countries except Senegal, the age at which people become eligible to receive benefits is higher than the average life expectancy.
2. The World Bank Pension Reform Options Simulation Toolkit (PROST) model assumes no change to the system and its governing parameters over the next 50 years. It should also be noted that this estimate does not include spending on military pensions, as data are not available to include this category in the forecasts.
3. This assumes perfect targeting of the programs. From 2015, the coverage of PSNP will increase to roughly 11 percent of the population.
4. Unless otherwise referenced, estimates of safety net coverage in East Africa are based on primary data collected expressly for this book from service providers (e.g., government agencies and aid agencies). Estimates of safety net coverage in the Sahelian countries come from World Bank social safety net studies, as well as from updated data provided by the World Food Programme and the United Nations Children's Fund (UNICEF).
5. The government of Senegal is the first in the region to have measured and acknowledged the inefficiency of universal subsidies. Senegalese officials agree with aid partners that a better system of targeted safety nets will be more efficient in addressing vulnerabilities. Before universal subsidies are phased out and replaced by cash transfers, however, efforts will be needed to improve the performance of markets for fuel and imported staples (see World Bank 2013).
6. Even so, for some households, depending on the context, this process can take a long time.
7. The costs of not protecting poor populations from the negative effects of shocks are high and long lasting. Ethiopian households that suffered during the 1984–85 drought continued to experience 2–3 percent less annual per capita growth in the 1990s. Children in households in Burkina Faso that experience a negative income shock are less likely than other children to enroll in school. The negative consequences of reducing investments in children can be irreversible: malnutrition alone lowers GDP growth by 2–3 percent.
8. The cost of basic coverage estimated in the umbrella model is equal to US$261 per household, which includes a blend of cash transfers, cash for work, and insurance subsidies. A per capita transfer of US$60 includes a 15 percent administration fee and 20 percent leakage to non-poor households.

## References

Hoddinott, J., and J. Lind. 2013. "The Implementation of the Productive Safety Nets Programme in Afar and Somali Regions, Ethiopia: Lowlands Programme Outcomes Report." International Food Policy Research Institute, Washington, DC.

Lind, J., and S. Kohnstamm. 2014. "Review of Social Protection Programmes and Projects in the IGAD Region." Background report prepared for the World Bank. World Bank, Washington, DC.

Monchuk, V. 2014 "Reducing Poverty and Investing in People: The New Role of Safety Nets in Africa. Experiences from 22 Countries." World Bank, Washington, DC.

UNOCHA (UN Office for the Coordination of Humanitarian Affairs). 2015. "Sahel: Food Insecurity 2011–2015." UNOCHA, New York/Geneva. ReliefWeb. http://reliefweb. int/disaster/ot-2011-000205-ner.

Venton, C.C., C. Fitzgibbon, T. Shitarek, L. Coulter, and O. Dooley. 2012. "The Economics of Early Response and Disaster Resilience: Lessons from Kenya and Ethiopia." Economics of Resilience Final Report. https://www.gov.uk/government/uploads/ system/uploads/attachment_data/file/67330/Econ-Ear-Rec-Res-Full-Report_20.pdf.

World Bank. 2010. "Designing and Implementing a Safety Net in Low-Income Setting: Lessons Learned from Ethiopia's Productive Safety Net Program." World Bank, Washington, DC.

———. 2012. "Managing Risk, Promoting Growth: Developing Systems for Social Protection in Africa." In the *World Bank's Africa Social Protection Strategy 2012–2022,* pp. 77. Human Development Africa. Washington, DC: World Bank.

———. 2013. "Republic of Senegal—Senegal SP—Safety Net Assessment." Africa Social Protection Group for West and Central Africa (AFTSW). World Bank, Washington, DC.

———. 2014. "Productive Safety Nets Project 4. Project Appraisal Document." World Bank, Washington, DC.

## Chapter 11

# Disaster Risk Management: Being Prepared for Unforeseen Shocks

*Carl Christian Dingel, Christoph Putsch, Vladimir Tsirkunov, Jean Baptiste Migraine, Julie Dana, Felix Lung*

## Current situation

Dryland regions of Sub-Saharan Africa are frequently hit by extreme weather and climate events, notably droughts and floods. Between 1970 and 2014, around 1,300 disasters (that is, drought, floods, storms, extreme temperatures landslides, volcanoes, and earthquakes) were reported across the region, of which approximately 40 percent occurred in 17 countries that have a predominantly dryland character[1] (EMDAT/Guha-Sapir, Below, and Hoyois 2015). In many of these dryland countries—particularly those located in the Sahel and the Horn of Africa—a large portion of the population is exposed to multiple hazards (Dilley et al. 2005).

The population in drylands disproportionately suffers from disasters. Although dryland countries contain only about one-third of the population in Sub-Saharan Africa, they account for more than 50 percent of those affected by disasters and nearly 80 percent of all casualties from disasters. The disproportionately large casualties suffered by dryland countries can be attributed to the fact that these countries contain large numbers of vulnerable people who are chronically exposed to drought, combined with the fact that many of these countries have limited capacity to prepare effectively for unforeseen shocks. Droughts have been responsible for the largest number of people affected by far, but during the past two decades the number of flood events has increased noticeably (table 11.1).

Globally droughts and floods are high-frequency and relatively low-severity events, in comparison to earthquakes and cyclones. Droughts are slow- or delayed-onset events, usually stretching over several years and exacting high economic losses, but causing limited infrastructure damage. Floods tend to be fast-onset disasters, often causing substantial infrastructure damage. In many dryland countries of Africa, a large proportion of the population is at risk from two or

**Table 11.1** Number of disaster events, people affected, and casualties per disaster type, Sub-Saharan Africa and dryland countries, 1970–2014

|  | Population 2013 |  | Drought | Flood | Storm | Earthquake | Volcano |
|---|---|---|---|---|---|---|---|
| **Sub-Saharan Africa (n = 49)** | 938 million | **Events** | 232 | 512 | 163 | 29 | 15 |
|  |  | **Affected** | 332.2 m | 58.3 m | 16.1 m | 0.5 m | 0.3 m |
|  |  | **Deaths** | 545,081 | 16,496 | 4,404 | 2,201 | 786 |
| **Predominantly dryland countries (n = 17)** | 293 million | **Events** | 114 | 194 | 19 | 3 | 4 |
|  |  | **Affected** | 197.4 m | 15.2 m | 0.3 m | 0.1 m | > 0.1 m |
|  |  | **Deaths** | 443,186 | 7,585 | 517 | 299 | 69 |

*Source:* EMDAT 2015.

*Note:* Predominantly dryland countries include Botswana, Burkina Faso, Cape Verde, Chad, Djibouti, Ethiopia, Eritrea, The Gambia, Kenya, Mali, Mauritania, Namibia, Niger, Senegal, Somalia, Sudan, South Sudan. m = million.

more hazards (usually drought and floods), including Niger (76 percent), Ethiopia (69 percent), Kenya (63 percent), and Burkina Faso (63 percent) (World Bank 2006). Projections for 2030 indicate that these countries will have very high levels of vulnerability to disasters and poverty (Shepherd et al. 2013).

## Impacts of disasters

Natural disasters impose a large financial burden on governments in two ways: (1) by causing immediate economic losses, and (2) by forcing resources to be redirected to short-term humanitarian assistance and away from longer-term development activities. The costs of disasters show up clearly in macroeconomic indicators, both in the short run as GDP losses and over the longer term as lasting declines in GDP growth. Total economic losses caused by disasters are modest in Sub-Saharan Africa relative to other regions, but when they are considered taking into account the size of the economies of many African countries and the fiscal budgets, the financial impact of disasters is extremely high. Furthermore, macroeconomic indicators do not always reflect the pain and suffering caused by disasters at the micro level: extreme weather and climate events disproportionately affect the poor, meaning that disasters tend to have severe impact on the livelihoods of the most vulnerable households and have the effect of pushing additional people into poverty. As discussed earlier, in dryland regions of Africa many of those affected by extreme weather events (figure 11.1) are often poor livestock-keepers and farmers (table 11.2).

Table 11.3 presents data on the damage and losses suffered in several dryland countries in which the World Bank has recently supported Post Disaster Needs Assessments (PDNAs). In Kenya, for example, the overall damage from the 2008–11 drought was estimated at US$12.1 billion (Government of Kenya 2012), with the majority (72 percent) of the losses falling on individuals, households, or businesses owning livestock, mostly in the northern drylands. During this period

**Figure 11.1** Population affected by droughts, floods, storms, earthquakes, and volcanoes in dryland countries, 1970–2013 (millions)

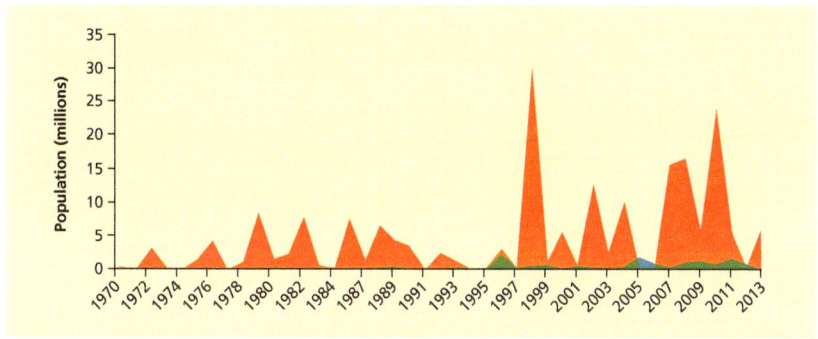

*Source:* EMDAT 2015.

*Note:* Red areas in the figure represent the number of people affected by drought; blue areas represent the number of people affected by floods.

**Table 11.2** Direct and indirect financial impact of natural disasters on different groups

|  | Government | Farmers/Herders | The Poorest |
|---|---|---|---|
| Direct Impacts | • Emergency response and recovery expenditures<br>• Expenditure on social and economic recovery programs<br>• Realization of contingent liabilities to state-owned enterprises, to firms critical to economic recovery | • Reconstruction costs for mostly uninsured assets<br>• Restocking/replanting/rehabilitation of productive assets | • Reconstruction costs for damaged assets<br>• Replacement of livestock |
| Indirect Impacts | • Decreased tax revenue due to economic disruption and declines in GDP growth<br>• Opportunity cost of diverting funds to disaster response and reconstruction<br>• Increased domestic/international borrowing costs<br>• Potential negative impact on sovereign credit ratings<br>• Increased expenditures for social support programs (safety nets)<br>• Migration due to disaster impact | • Loss of income due to interruption of crop/livestock/fish stock production<br>• Loss of income due to economic decline and/or lack of access to markets<br>• Increased borrowing costs<br>• Increased risk aversion to new and innovative investments | • Decreases in expenditure on food, accommodation, and human capital<br>• Loss of social support due to breakdown in informal safety net systems such as family support<br>• Loss of income and employment<br>• Increased borrowing costs |

*Source:* Calculated based on data published in the Post Disaster Needs Assessments, various countries.

average GDP was reduced by 2.8 percent/year. In addition, from 2007–08 until 2010–11, humanitarian relief expenditure of the Government of Kenya rose to US$125 million/year, complemented by US$241 million/year from international donors (up from decade averages of US$57 million/year and US$102.2 million/year, respectively) (Government of Kenya 2012).

**Table 11.3** An overview of damages and losses from recent PDNAs, selected dryland countries

| Event | Country | Year | People affected (thousands) | Damage (US$ million) | Losses (US$ million) | Recovery cost (US$ million) |
|---|---|---|---|---|---|---|
| Floods | Namibia | 2009 | 350 | 136 | 78 | 622 |
| Floods | Burkina Faso | 2009 | 150 | 102 | 33 | 266 |
| Floods | Senegal | 2009 | 485 | 56 | 48 | 204 |
| Drought | Kenya | 2008–11 | 3,700 | 8 | 11,300 | 17,700 |
| Floods | Sudan | 2013 | 340 | 134 | | |

*Source:* Calculated based on World Bank data.
*Note:* PDNA = Post Disaster Needs Assessment.

## Limited capacity to manage unforeseen shocks

The capacity of countries to reduce disaster risk and prepare for unforeseen extreme weather and climate events is limited in many parts of Sub-Saharan Africa. Shepherd et al. (2013) summarized the disaster risk management capacity of countries in a composite score designed to capture the capacity of each country to prevent disasters from causing impacts, now and in the future. Many dryland countries fare poorly according to this scale: *Very poor:* Chad, Sudan, and Somalia; *Poor:* Niger and Mauritania; *Average:* Ethiopia, Kenya, and Mali; *Better than average:* Burkina Faso and Senegal; and *Relatively good:* Namibia and Botswana.

Following the severe droughts and resulting food crises of the 1960s, 1970s, and 1980s, many dryland countries in East and West Africa started to collaborate on monitoring and managing drought and food security. Two regional organizations took the lead in these functions: the Permanent Interstate Committee for Drought Control in the Sahel (CILSS), established in West Africa in 1973, and the Intergovernmental Authority on Development (IGAD), established in the Horn of Africa in 1983. Since then, regional policies, operational monitoring frameworks, and systems to improve regional climate projections have been instituted. In 2006 the Economic Community of West African States (ECOWAS) formulated a regional disaster risk reduction policy (ECOWAS 2006), and in 2010 the African Union together with the World Food Programme initiated the African Risk Capacity (ARC) as a regional climate insurance mechanism (see box 11.1).

Despite these advances, however, no harmonized framework is as yet in place for monitoring and response that integrates multiple hazards—droughts, floods, food security, and other hazards, including locusts, extreme temperatures, and fires. The underlying data and hydrometeorological and production forecasts remain weak, and capacity to deal with rapid onset hazards is yet to be developed. In some countries flood forecasting systems have been put in place, but they tend to have a local focus.

BOX 11.1

# African Risk Capacity (ARC)

African Risk Capacity (ARC) is a Specialized Agency of the African Union.[a] ARC's mission is to help African Union (AU) member states improve their capacities to better plan, prepare, and respond to extreme weather events and natural disasters and to assist those affected in a timely and effective manner. As a continental sovereign risk pool, ARC provides cost-effective contingency funding to African governments to execute preapproved contingency plans should severe events occur. Developed as a joint project of the AU and the United Nations World Food Programme (WFP), the ARC became a Specialized Agency of the AU in November 2012. It currently counts 25 AU countries as members and is supervised by a governing board of African ministers and experts chaired by Nigeria's Coordinating Minister for the Economy, Dr. Ngozi Okonjo-Iweala.[b]

While the ARC agency provides member states capacity-building services for insurance, contingency planning, and operations, a nationally regulated financial affiliate, the ARC Insurance Company Limited (ARC Ltd), was established to execute the risk transfer operations. ARC Ltd was registered in Bermuda in December 2013 and started operations in 2014. A specialist hybrid mutual insurance company and Africa's first-ever disaster insurance pool, ARC Ltd aggregates risk by issuing insurance policies to participating governments and transferring some of that risk to the international market. It uses the satellite weather surveillance software *Africa RiskView*, developed by WFP, to estimate the impact of drought on vulnerable populations—and the response costs required to assist them—before a season begins and as it progresses, so that index-based insurance payouts, based on *Africa RiskView*, are triggered at or before harvest time if the rains are poor. With a US$200 million initial capital commitment provided by the governmental development agencies of Germany (KfW) and the United Kingdom (DFID), ARC Ltd issued drought insurance policies totaling US$129 million for a total premium cost of US$17 million to a first group of African governments for five rainfall seasons—Kenya, Mauritania, Niger, and Senegal—in May 2014, marking the launch of the inaugural ARC pool. Seven additional countries are in the queue to join the next pool in 2015, with a target of up to 20 countries receiving coverage for drought, flood, and cyclones totaling over US$600 million in the next five years. In addition to insurance for weather events, ARC has recently been mandated by its member states to develop coverage for disease outbreaks and epidemics, such as Ebola, and is developing a climate change adaption financing mechanism for its insured countries to respond to the impacts of increased climate volatility.

*(continued next page)*

**Box 11.1** *(continued)*

In January 2015, ARC issued payouts totaling US$25 million to Senegal, Mauritania, and Niger as a result of drought conditions during the 2014 rainfall season to implement pre-prepared response plans to assist affected populations. These payouts were triggered before the UN Sahel Humanitarian Appeal in February 2015 and will be used to implement targeted food distributions, subsidized fodder sales and scale cash transfer, and school feeding programs in the recipient countries.

*Notes:*

a. http://www.africanriskcapacity.org

b. WFP continues to provide administrative services support to the ARC agency, including procurement and trust fund management, through an administrative services agreement.

Reliable hydrometeorological systems and services are needed to ensure timely early warning and preparedness, yet such systems and services are lacking in many dryland countries. A recent survey carried out by the World Meteorological Organization (WMO) concluded that in Sub-Saharan Africa "there are wide-spread deficiencies in hydrometeorological observation networks, tele-communications, and informatics systems … and very limited capacity in data management and product customization. The national hazard warning capacities are uneven, even nonexistent in some countries, while warning programs often do not address all significant meteorological and hydrological hazards." According to WMO standards, Sub-Saharan Africa ranks last among all regions in terms of land-based observation networks, meeting only about one-eighth of the minimum requirements (Rogers and Tsirkunov 2013).

Even if national policies for disaster response can be strengthened, and even if substantial supporting investments can be made in resilience-building mechanisms, it is likely that livelihoods and economies throughout the region will continue to be adversely affected by droughts and floods. This in turn means that governments will continue to be exposed to disaster-linked expenses, such as the costs of mounting humanitarian responses when disasters strike. To cover such costs, most African countries have historically relied on funding mobilized post-disaster, such as loans or donor assistance. Figure 11.2 shows the trend of humanitarian assistance spending for crises, conflicts, and disasters provided by donors between 2000 and 2011 to predominately dryland countries in Africa (GHA 2015), along with the number of people affected by droughts and floods. While post-disaster financing can sometimes be accessed on more favorable terms than prearranged financing, it may take a long time to negotiate (e.g., emergency loans) or turn out to be highly unpredictable (e.g., donor assistance), with the result that development programs that may have been under implementation for many years can end up being threatened.

**Figure 11.2** People affected by droughts and floods in dryland countries and costs of humanitarian interventions, 2000–11

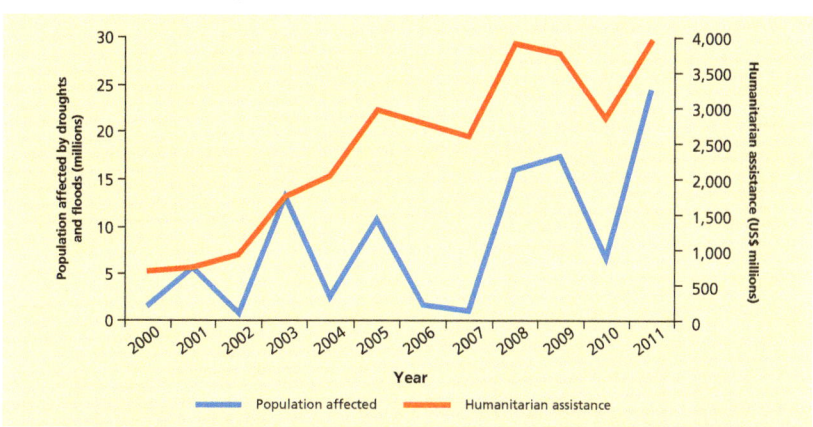

*Source:* EMDAT 2015; GHA 2015.

*Note:* The figure uses data for Botswana, Burkina Faso, Cape Verde, Chad, Djibouti, Eritrea, Ethiopia, The Gambia, Kenya, Mali, Mauritania, Namibia, Niger, Senegal, Somalia, South Sudan, and Sudan.

One strategy for reducing the uncertainties associated with relying on post-disaster funding is to institute risk financing instruments that can be established before a disaster hits, such as insurance against disasters or reserve funds. Such instruments allow governments to shoulder the financial burden of dealing with disasters before they occur. Risk financing instruments that can be established before a disaster hits can avoid certain drawbacks, but they require considerable advance planning, they can be expensive, and they may be limited in scope (Mahul and Cummins 2009). An additional problem is that the types of risk-financing instruments of greatest relevance to disasters (e.g., weather-indexed agricultural insurance) are still in their infancy in many dryland countries and hence are poorly understood.

## Opportunities

Disaster risk management strategies and programs, when correctly designed and effectively implemented, can play an important role in reducing vulnerability and increasing resilience of people living in drylands. They can do this by reducing exposure to shocks of vulnerable households, reducing sensitivity to shocks, and especially by improving the capacity of shock-affected households to cope after a shock has occurred.

### Reducing sensitivity

Disaster risk management focuses on reducing risks and better preparing for extreme weather and climate events (that is, better managing any residual risk).

Risk reduction includes (1) reducing vulnerability, (2) better understanding hazards, and (3) managing exposure. A strong disaster risk management capacity is essential to minimize the potential for long-term losses resulting from the impacts of hazards on vulnerable, exposed people (Shepherd et al. 2013).

**Early warning systems and hydromet services.** Early warnings, climate outlooks for the rainy season, and extreme weather event forecasts are important elements for reducing exposure and sensitivity to extreme weather and climate events in drylands. Following the droughts of the 1960s, 1970s, and 1980s, the WMO regional climate centers—the African Center for Meteorological Applications for Development (ACMAD), the IGAD Climate Prediction and Applications Center (ICPAC), and the SADC Climate Service Center (SADC CSC)—successfully initiated seasonal climate outlook forums for West Africa (PRESAO),[2] the Greater Horn of Africa (GHACOF),[3] and Southern Africa (SARCOF).[4] Climate outlook forums provide consensus-based, region-specific, seasonal climate forecasts for upcoming cropping cycles and have become an important instrument for understanding the weather risks in drylands (WMO 2009). Pilot projects have demonstrated the ability to help farmers who formerly relied on seasonal forecasts to reduce the sensitivity of their cropping activities to extreme climate events, thereby raising or at least stabilizing their agricultural income. Benefits realized by vulnerable farmers include better planning, more efficient utilization of inputs, avoidance of crop damage from extreme weather events and pests, and better management of stocks (Archer et al. 2007; WMO 2005). Humanitarian agencies have taken notice of these gains and are increasingly using climate outlook information to plan interventions and take early actions, such as stocking up goods for relief operations (Coughlan de Perez and Mason 2014).

The WMO regional climate centers are working with national hydrometeorological services to link local observation capacity with global and regional weather prediction models. By mobilizing a wide range of actors at many different levels, they can forecast severe weather using a cascading approach in which information passes from regional centers to national hydrometeorological services. In this way the global products of the major numerical prediction centers can be used by even the most capacity-limited national hydrometeorological services; these services can use the information to improve their alerting and warning services while avoiding the cost of stand-alone investments in high-end computing infrastructure and staffing. The WMO Severe Weather Forecasting Demonstration Project has successfully piloted this approach, increasing the lead time and reliability of alerts related to high-impact events, such as heavy precipitation and severe winds (WMO 2010).

The potential benefits from establishing effective hydrometeorological systems and early warning capacity are often underestimated. Hallegatte (2012) estimates that upgrading all hydrometeorological information systems and early-warning capacity in developing countries worldwide would prevent between US$300 million and US$2 billion annually in disaster-related losses, save 23,000 lives annually on average, and provide between US$3 billion and US$30 billion in additional economic benefits. Studies carried out in Switzerland and the United States show high returns to investments in improved meteorological and hydrological services, with cost-benefit ratios ranging from 1:4 to 1:6 (Rogers and Tsirkunov 2013). A recent World Bank study focusing on Europe and Central Asia generated similar results, reporting estimated cost-benefit ratios ranging from 1:2 to 1:10 (Tsirkunov et al. 2007). The benefits of high-priority hydrometeorological investments in Africa alone would likely exceed US$1 billion over the next 10–15 years (World Bank 2014).

## Improving coping capacity

Being well prepared can reduce exposure and sensitivity to shocks, but it will not eliminate vulnerability completely. Even if substantial investments are made in risk reduction mechanisms, disasters will likely continue to occur in the drylands, with adverse impacts on the livelihoods of people who live there. Preparing for the unforeseen is therefore important, so that instruments will be available to help disaster-affected households cope after a disaster has occurred. Measures to strengthen coping capacity include: (1) pooling, transferring, and sharing risks; (2) effectively preparing for extreme events; and (3) managing resilient recovery and reconstruction.

**Risk finance and insurance to pool and transfer risk.** Governments can take steps to reduce the negative financial effects of disasters in a way that protects both people and assets. This requires short-, medium-, and long-term policy interventions focused equally on risk reduction and on financial risk management. Disaster risk financing and insurance solutions can help countries minimize the cost and optimize the timing of post-disaster funding needs without compromising development goals, fiscal stability, and well-being. Disaster risk financing and insurance therefore must be an integral part of the disaster and climate risk management agenda in dryland regions. Disaster risk financing and insurance complement disaster risk management activities by securing adequate financial resources to cover residual risks that cannot be mitigated and by creating the right financial incentives to invest in risk reduction and prevention.

Instruments that offer the greatest potential benefits for dryland countries can be grouped into three categories:

- **Sovereign disaster risk financing** aims to increase the capacity of national and subnational governments to provide immediate emergency funding as well as long-term funding for reconstruction and development.
- **Agricultural insurance** aims to protect farmers, herders, and fishermen from loss arising from damage to their productive assets. For example, the

Government of Kenya is in the process of establishing a national agricultural insurance scheme for agricultural producers against drought.

- **Disaster-linked social protection** helps governments strengthen the resilience of the poorest and most vulnerable to the debilitating effects of natural disasters. It does this by applying insurance principles and tools to enable social protection programs such as social safety nets to scale up and scale out assistance to beneficiaries immediately following disaster shocks.

While governments may not need to pursue all of these policy options, a comprehensive disaster risk financing strategy typically would build on some combination of them. Together they help to clarify, reduce, and manage public contingent liabilities to natural disasters, thereby making financing of disaster-linked expenses more cost-effective, timely, and reliable. The elements of such a strategy are as follows:

- **Cost-effectiveness.** The more quickly financing can be made available following the onset of a disaster, the more costs can be contained. In the case of droughts, losses continue over time, extending far beyond short-run agricultural production losses. They can include, for example, loss of productive assets, reduced food consumption, lower rates of educational enrolment, higher incidence of disease, and ultimately, loss of lives.

- **Timeliness.** Rapid mobilization of funds to support scalability in response to drought is crucial to limit the negative impacts that a population experiences and contain contingent liabilities to finance the required relief efforts. An early, well-targeted response—for example through the scaling up of a social safety net—can cost a fraction of the emergency aid required after a famine evolves.

- **Reliability.** Clear, predefined rules of disbursement, as typically encountered in risk financing instruments, can make the provision of needed funds more predictable and more dependable. Insurance policies that emit payments based on objective and easily measurable rules can make budgetary planning much easier for government agencies and relief organizations.

A comprehensive ex ante financing strategy, as depicted in figure 11.3, involves layering various types of risk, where the "lower layers" refer to more frequently occurring, low-impact disaster events, while the "upper layers" represent infrequent but more extreme events. Depending on the layer, different risk financing mechanisms tend to be most cost-effective. For example, for lower-layer natural disasters, reserve funds often present the most cost-effective solution (Mahul and Cummins 2009).

Governments in Sub-Saharan Africa, similarly to governments in other developing regions, often rely on ad hoc measures to respond to the incremental financing needs that arise in the case of disasters, including dipping into disaster

**Figure 11.3** Catastrophic risk layering and respective cost-effective risk financing

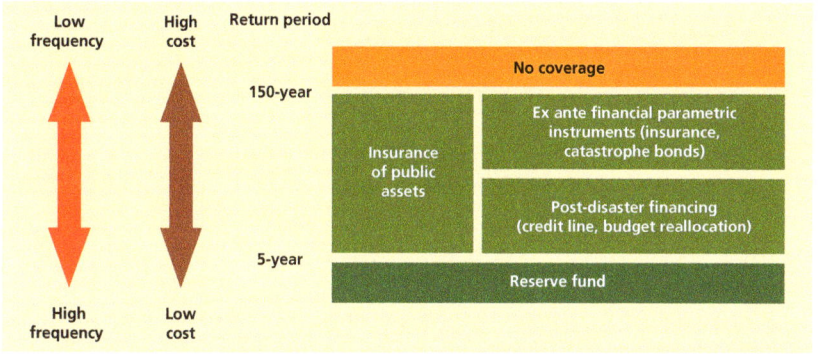

*Source:* Cummins and Mahul 2009.

funds, undertaking emergency budget reallocations, and launching urgent appeals for donor assistance. Such measures tend to be unreliable and often inadequate. In many countries, therefore, a shift is taking place, and efforts are being made to draw on a wider range of financial instruments, including many that can be lined up in advance. The policy objective being pursued by these countries is to build financial resilience at the national level, which requires paying attention to a series of considerations as follows:

- **Appropriate risk information.** Appropriate risk information allows public and private decision makers to assess the underlying price of risk and clarify costs and benefits of investing in risk reduction or risk financing.

- **Ownership of risk.** Clarifying who is responsible for a certain risk, establishing the contingent liability of the governments, donors, private sector, and other groups is an important starting point. Furthermore, clearly established rules under social protection programs give predictability to vulnerable populations and enable better planning and budgeting.

- **Cost of capital.** Access to capital at different costs is necessary for effective emergency response, reconstruction, and risk reduction and prevention.

- **Timeliness.** Different types of funds need to be available at the appropriate time following a disaster to cover relief, response, and reconstruction efforts. The rapid mobilization of funds to support relief efforts is crucial to limit humanitarian costs.

- **Discipline.** Disaster risk financing helps affected groups plan in advance of a disaster and agree beforehand on rules and processes for securing funds (budget mobilization) and spending funds (budget execution). This creates greater discipline, transparency, and accountability in post-disaster spending.

**Insurance for farmers, herders, and fishermen.** Agricultural insurance can protect farmers, herders, and fishermen from losses arising from damage to productive assets. This helps farmers and herders increase their awareness and understanding of financial vulnerability to agricultural risks; to possibly adopt riskier, but higher yielding production methods; and to have in general a better understanding of the financial services suitable for low-income households.

Responding to recurring extreme drought events in northern Kenya, the Kenyan Ministry of Agriculture, Livestock, and Fisheries (MALF) recently conducted a diagnostic study investigating options for large-scale crop and livestock insurance in Kenya. For livestock, the study proposed insurance based on a satellite-based index of ground cover. The objective of the insurance would be to provide asset protection through a policy that provides financial compensation in times of severe drought. Payouts could be used to mobilize access to fodder and other life-saving services.

### Resilient recovery

In considering how to improve coping capacity, it is important to highlight the importance of building back better so as to increase resilience in the future. In addition to quantifying losses and assessing damage to the economy, many PDNAs assess the needs for reconstruction and recovery. By making concrete recommendations for building back better and strengthening disaster risk management, they can lay out a roadmap for strengthening resilience in the future. Despite their adverse impacts, disasters can create opportunities for planning and rebuilding more resilient livelihoods and economies.

PDNAs are anchored by the joint declaration on Post-Crisis Assessments and Recovery Planning—a 2008 tripartite agreement between the EU, United Nations Development Programme (UNDP), and the World Bank to coordinate all post-crisis interventions under the lead of the affected country's government (EU, UNDG, and World Bank 2008). The PDNA process is complemented by comprehensive, integrated recovery plans that prioritize and sequence recovery interventions and help governments improve their readiness for future disasters, as outlined by the Global Facility for Disaster Reduction and Recovery (GFDRR 2015). In this way the international community can provide support for transitioning swiftly from response to recovery and reconstruction.

In summary, a comprehensive disaster risk management framework must include measures designed to reduce risks by limiting exposure and sensitivity to shocks before they occur, as well as measures designed to manage residual risks and improve coping capacity after a shock has occurred. The complementary relationships between these two sets of measures are illustrated in figure 11.4.

**Figure 11.4** Disaster risk management framework

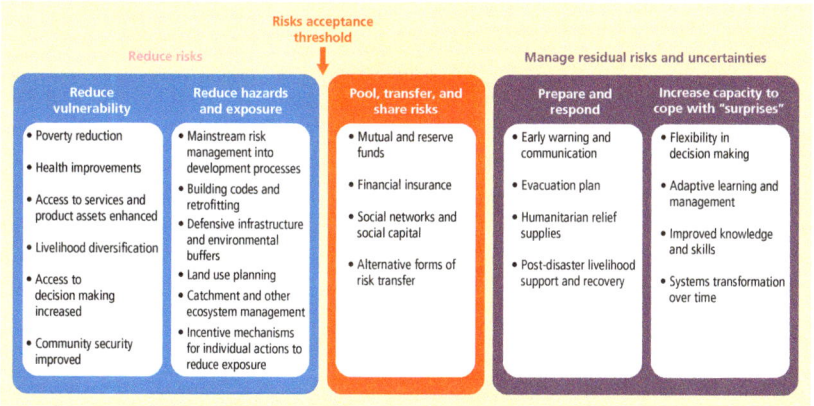

*Source:* Adapted from IPCC 2012.

## Challenges

Efforts to reduce risk and prepare for unforeseen weather and climate shocks are challenged in many dryland countries by the use of inappropriate instruments, capacity constraints, lack of coordination, resource limitations, lack of political leadership, and underdeveloped markets.

**Limited use and application of information.** Interventions to manage risk often end up failing to address the particular circumstances or dynamics of poverty and vulnerability of the drylands. For example, weather and climate information should be provided in a way that is meaningful to farmers and livestock herders, but also to women and other user groups, and it should enable them to take appropriate action. Similarly, early warning systems are only as good as the action that follows the warning. Contingency plans, interventions, and recovery activities need to be tailored to different vulnerable groups and to take their specific livelihoods into account (e.g., the semi-sedentary livelihood of herders).

**Lack of coordination and integration.** Climate forecasting in dryland regions and early warning mechanisms so far have focused on drought and related food security issues, but the systems currently in place lack an integrated (multi-hazard) and regionally harmonized approach for effective early warning.

**Regional capacity constraints.** The national hydrological and meteorological services would benefit from more coordinated support from global, regional, and subregional centers to better use resources available within a cascading process and to increase the benefits to end users, that is, people and economies. A recent institutional assessment of ACMAD (African Centre of Meteorological Applications for Development) and AGRHYMET[5] (a drought-monitoring and capacity-building center in the West Africa Region) showed that even though these regional technical centers provide essential services, their ability to deliver is severely constrained by inadequate budgets, the lack of qualified staff, and weak infrastructure.

**Resource limitations.** Weather and climate services in the drylands remain largely underfunded. Even though substantial investments have been made in some cases in state-of-the-art infrastructure, additional resources need to be committed to cover operating costs, which typically run on the order of 10–15 percent of the investment costs annually (Rogers and Tsirkunov 2013). Without operating budgets, many national hydrometeorological services lack the capacity to conduct effective forecasting for extreme events and to communicate the information in a timely manner.

**Lack of political leadership.** Courage is required on the part of governments and traditional aid agencies to move away from a system funded through international appeals and toward a system based on ex ante funding that incorporates mechanisms for transferring and pooling risk.

**Underdeveloped insurance markets.** The insurance market for vulnerable groups, notably farmers and herders, remains underdeveloped. Although weather-indexed insurance has the potential to provide much needed protection to keep farmers out of extreme poverty and to support the ability to make investments in the future, pilot insurance schemes have often failed.

**Repeating vulnerabilities.** Recovery and reconstruction efforts often do not pay sufficient attention to underlying vulnerabilities and exposure to risk. Communities often rebuild their houses and assets with the same materials, move to other vulnerable areas, and lack the resources, for example, to retrofit their houses to make them more flood-resilient or manage their livestock herds better following a drought.

## Key messages

Disaster risk management can play a key role in strengthening resilience and reducing vulnerability in the drylands. Initiating a disaster risk management strategy that will be effective in the drylands requires action along the following lines.

**Investing in knowledge.** Understanding hazards and effective preparedness to extreme weather and climate events are the basis for most decisions on reducing risk and preparing for disasters. This requires more accurate forecasts and better weather, climate, and hydrological services. Hallegatte (2012) recommends investment in five domains: (1) local observation systems; (2) local forecast capacity; (3) increased capacity to interpret forecasts and translate them into warnings; (4) communication tools to distribute and disseminate information, data, and warnings; and (5) institutional capacity building and increased decision-making capacity by the users of warnings and hydrometeorological information.

**Paying attention to local circumstances.** The design and delivery of disaster risk management interventions should be tailored to the livelihoods of the drylands and should integrate food security and other disasters in an effective manner. Forecasting and early warning should build upon existing experience and institutions related to food security monitoring and integrate droughts, floods, and other disasters into disaster risk management strategies.

**Strengthening regional institutions.** Scaling up projects and tailoring national programs to deliver effective support in drylands requires creative responses to limited delivery capacity in many of these areas. Regional organizations play an important role in the Sahel, Horn of Africa, and Southern Africa as "knowledge centers" by facilitating data exchange, coordinating responses, and building capacities of member states. There is a need to identify sustainable financing solutions for these institutions.

**Empowering national governments.** Financial protection requires strong leadership by a country's ministry of finance. Strong stewardship is required, as disaster risk financing brings together disaster risk management, fiscal risk and budget management, public finance, private sector development, and social protection. Disaster risk financing and insurance is a long-term agenda that requires political will, technical expertise, and time. While simple measures can quickly support improved financial protection, more complex financial solutions and institutional change require technical expertise and political will.

**Mobilizing the capacity of the private sector.** The private sector is an essential partner in disaster risk management. It can bring capital, technical expertise, and innovative financial solutions to better protect the government and society against natural disasters.

**Building back better.** Disasters present opportunities for engagement on risk reduction, and this should be reflected in all post-disaster engagement. The aftermath of a disaster often focuses resources and political will on reducing existing risk and preventing future risk, creating opportunities to "build back better" and begin systematic engagements.

## Notes

1. These countries include Botswana, Burkina Faso, Cape Verde, Chad, Djibouti, Ethiopia, Eritrea, The Gambia, Kenya, Mali, Mauritania, Namibia, Niger, Senegal, Somalia, South Sudan, and Sudan.
2. PRESAO – Previsions Saisionnaieres en Afrique de l'Ouest
3. GHACOF – Greater Horn of Africa Climate Outlook Forum
4. SARCOF – Southern African Regional Climate Outlook Forum
5. AGRHYMET = AGRrometeorology, HYdrology, METeorology

## References

Archer, E., E. Mukhala, S. Walker, M. Dilley, and K. Masamvu. 2007. "Sustaining Agricultural Production and Food Security in Southern Africa: An Improved Role for Climate Prediction?" *Climatic Change* 83: 287–300.

Coughlan de Perez, E. and S. J. Mason. 2014. "Climate Information for Humanitarian Agencies: Some Basic Principles." *Earth Perspectives* 1(11).

Cummins, D., and O. Mahul. 2009. "Catastrophe Risk Financing in Developing Countries: Principles for Public Intervention." World Bank, Washington, DC.

Dilley, M., R.S. Chen, U. Deichmann, A.L. Lerner-Lam, M. Arnold, J. Agwe, P. Buys, O. Kjevstad, B. Lyon, and G. Yetman. 2005. "Natural Disaster Hotspots: A Global Risk Analysis." World Bank, Washington, DC. http://documents.worldbank.org/curated/en/2005/04/6433734/natural-disaster-hotspots-global-risk-analysis.

ECOWAS (Economic Community of West African States). 2006. "ECOWAS Policy for Disaster Risk Reduction." Humanitarian Affairs Department, Abuja, Nigeria.

EMDAT. 2015. Emergency Events Database, accessed January 2015. WHO Collaborating Centre for Research on the Epidemiology of Disasters (CRED), Universite Catholique Louvain, Belgium.

EMDAT/Guha-Sapir, D., R. Below, and P. Hoyois/EMDAT. 2015. International Disaster Database (www.emdat.be). Universite Catholique de Louvain, Brussels, Belgium.

EU (European Union), UNDG (United Nations Development Group), and World Bank. 2008. "Joint Declaration on Post-Crisis Assessments and Recovery Planning." Signed September 25, 2008. EU, UNDG, and World Bank.

GFDRR (Global Facility for Disaster Reduction and Recovery). 2015. "Guide to Developing Disaster Recovery Frameworks." Sendai Conference Version, March 2015. World Bank, Washington, DC.

GHA (Global Humanitarian Assistance). 2015. "Global Humanitarian Assistance Report 2015." GHA, Development Initiatives, U.K.

Government of Kenya. 2012. "Kenya Post-Disaster Needs Assessment (PDNA) 2008–2011 Drought." Prepared with support by the World Bank, United Nations, and European Union.

Hallegatte, S. 2012. "A Cost-Effective Solution to Reduce Disaster Losses in Developing Countries: Hydrometeorological Services, Early Warning, and Evacuation." Policy Research Working Paper 6058, World Bank, Washington, DC.

IPCC (Intergovernmental Panel on Climate Change). 2012. *Managing the Risks of Extreme Events and Disasters to Advance Climate Change Adaptation*. A Special Report of Working Groups I and II of the Intergovernmental Panel on Climate Change, eds. Field, C.B., V. Barros, T.F. Stocker, D. Qin, D.J. Dokken, K.L. Ebi, M.D. Mastrandrea, K.J. Mach, G.-K. Plattner, S.K. Allen, M. Tignor, and P.M. Midgley. Cambridge, UK and New York, NY: Cambridge University Press.

Mahul, O., and D. Cummins. 2009. "Catastrophe Risk Financing in Developing Countries." World Bank, Washington, DC.

Rogers, D.P., and V. Tsirkunov. 2013. *Weather and Climate Resilience: Effective Preparedness through National Meteorological and Hydrological Services.* Directions in Development: Environment and Sustainable Development. Washington DC: World Bank.

Shepherd, A., T. Mitchell, K. Lewis, A. Lenhardt, L. Jones, L. Scott, and R. Muir-Woods. 2013. "The Geography of Poverty, Disasters and Climate Extremes in 2030." Overseas Development Institute. London, UK.

Tsirkunov, V., S. Ulatov, M. Smetanina, and A. Korshunov. 2007. "Customizing Methods for Assessing Economic Benefits of Hydrometeorological Services and Modernization Programmes: Benchmarking and Sector-Specific Assessment." In *Elements for Life,* ed. Soobasschandra Chacowry. Geneva, Switzerland: World Meteorological Organization.

WMO (World Meteorological Organization). 2005. "Climate Outlook Forums and Agricultural Applications." WMO, Geneva.

———. 2008. *Capacity Assessment of National Meteorological and Hydrological Service in Support of Disaster Risk Reduction: Analysis of the 2006 WMO Disaster Risk Reduction Country-Level Survey.* Geneva: WMO.

———. 2009. "Regional Climate Outlook Forums." WMO, Geneva. www.wmo.int/ pages/prog/wcp/wcasp/documents/ RCOF_Flyer1.4_July2009_EN.pdf

———. 2010. "Severe Weather Forecasting Demonstration Project (SWFDP): Overall Project Plan, updated 2010." WMO, Geneva.

World Bank. 2012. "Managing Risk, Promoting Growth: Developing Systems for Social Protection in Africa." In the *World Bank's Africa Social Protection Strategy 2012–2022,* pp. 77. Human Development Africa. Washington, DC: World Bank.

———. 2013. "Ethiopia's Productive Safety Net Program (PSNP): Integrating Disaster and Climate Risk Management—Case Study." Working Paper 80622. World Bank, Washington, DC.

———. 2014. "Productive Safety Nets Project 4. Project Appraisal Document." World Bank, Washington, DC.

# Part C
# Toward Policy Priorities

Chapter **12**

# Evaluating Options: Assessing the Relative Merits of Resilience Interventions

*Raffaello Cervigni, Michael Morris, Federica Carfagna, Jawoo Koo, Joanna Syroka, Zhe Guo, Hua Xie, Balthazar de Brouwer, Elke Verbeeten*

## The scale of the development challenge

Returning to the projections presented in Chapter 4, it is useful to recap the scale of the challenge facing policy makers. The baseline projections generated under the business as usual (BAU) scenario of the numbers of vulnerable people likely to be living in the drylands of East and West Africa in 2030 provide a convenient yardstick that can be used to assess the attractiveness of the various interventions discussed in this book that are designed to improve resilience.

Across the 10 dryland countries for which sufficient data are available to allow modeling of resilience interventions, it is estimated that in 2010 about 30 percent of the population living in dryland zones was vulnerable to droughts and other shocks. While this number is quite large, fortunately not all vulnerable households experience a drought every year, and even those households that experience a drought do not necessarily see their income fall below the poverty line (the study's definition of "drought-affected"). Assuming historical climate patterns, the modeling simulations show that in any given year, approximately 20 percent of the vulnerable households are affected by drought, equivalent to about 6 percent of the total population in the 10 countries. Of course these are not the same people every year, as droughts occur in different locations and strike with different intensities.

The size of the drought-affected group is of interest because it determines the amount of resources to be mobilized every year—in the form of safety nets, international humanitarian assistance, or other forms of support—to assist people unable to cope with the effects of drought. The size of the drought-affected group also influences the mix of assistance that can be offered: for a given budget, the larger the group of drought-affected households, the larger the share of resources needed for short-term emergency response activities and the smaller the share of resources available to build longer-term resilience. Because it has important implications for policy making, the size of the drought-affected group is a key output of the umbrella model.

A second group of significance for the analysis consists of pastoralist households living in arid zones who own herds smaller than the minimum size needed to provide enough income to meet household consumption needs (estimated to be 5 Tropical Livestock Units [TLU] per household). For these households day-to-day survival appears extremely precarious, even in the absence of droughts and other shocks. This group—which accounts for 7 percent of the population across the entire sample of 10 countries but makes up a much larger share of the population in some countries, for example Niger (figure 12.1)—will come under increasing pressure to abandon pastoralism as its primary livelihood strategy and turn to other activities. In the umbrella model 2030 projections, it is assumed that pastoralist households owning fewer animals than the critical minimum level of 5 TLU will transition from pastoralism to crop farming.

In 2030 the umbrella model projects a 60 percent increase on average in the number of vulnerable people and a proportionally similar increase in the

**Figure 12.1** Share of 2010 pastoralist population likely to drop out of pastoralism by 2030 (%)

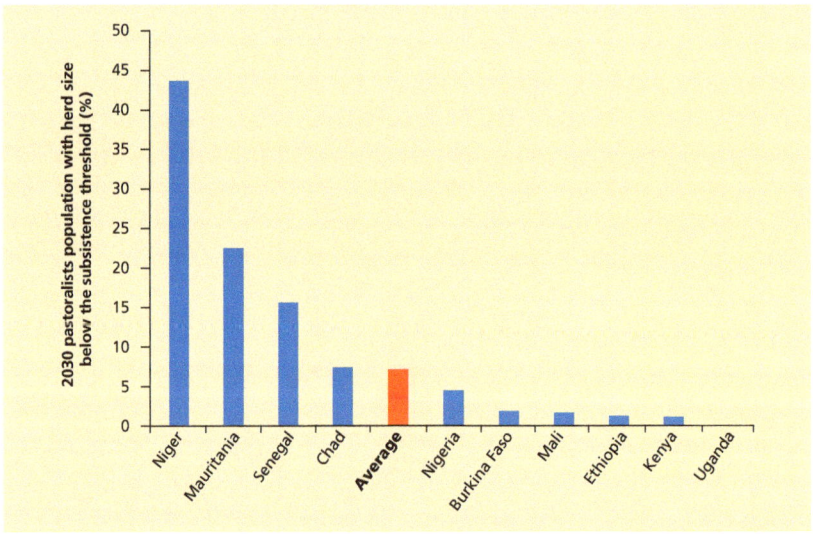

Source: De Haan 2016.

Note: The figures refer to pastoralists whose herds are below a minimum critical herd size; these pastoralists are expected to move out of pastoralism, and in the analysis are assumed to become farmers by 2030.

number of drought-affected people (figure 12.2). With the exception of Burkina Faso, by 2030 all of the countries in the sample are projected to experience increases in the number of vulnerable and drought-affected people. The increase is projected to be especially high in Niger, where the number of vulnerable and drought-affected people is expected to triple.

The projected increases by 2030 in the number of vulnerable and drought-affected people reflect the combined effects of several key drivers, including rapid population growth, relatively slow and inequitable economic growth, and binding bioclimatic and social constraints that limit the ability of the natural resource base to support greater numbers of animals. Most importantly, in pastoral areas prospects for expanding herd sizes at rates fast enough to keep pace with population growth are limited by the size of accessible grazing area.

To put the magnitude of the resulting challenge in perspective, the annual cost of bringing all drought-affected people up to the poverty line by providing support through social safety nets would range from 0.3 percent to almost 5 percent of GDP (figure 12.3). In interpreting these results it is important to keep in mind two points. First, these cost estimates are annual averages; in reality, financing needs would fluctuate dramatically and unpredictably, falling in years

**Figure 12.2** **People vulnerable/affected by drought in 2030 (2010=100)**

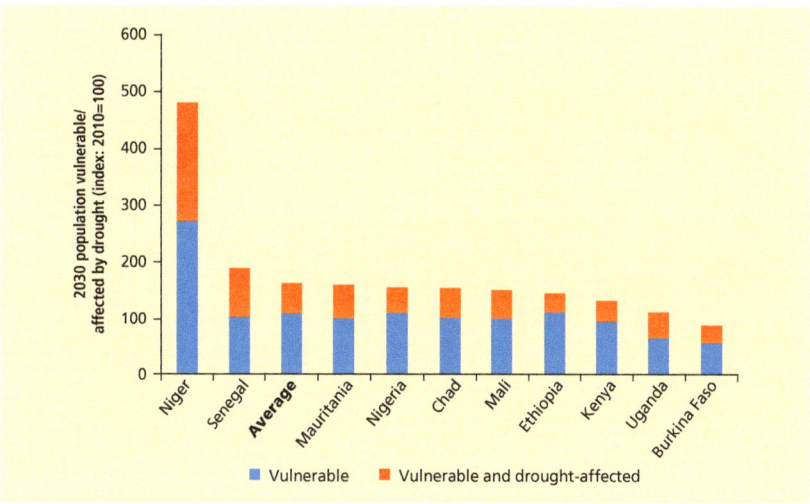

*Source:* Calculation based on the approach discussed in the Appendix.

*Note:* The total number of 2030 vulnerable people (proxied by the population living below the international poverty line) was estimated based on UN medium fertility population projections, as well as the average GDP growth scenario as defined in Carfagna, Cervigni, and Fallavier (2016). The average number of drought-affected people was estimated through the African Risk Capacity (ARC) model using crop yield simulations (see Appendix for details).

**Figure 12.3** Share of 2030 GDP required to protect drought-affected population (%)

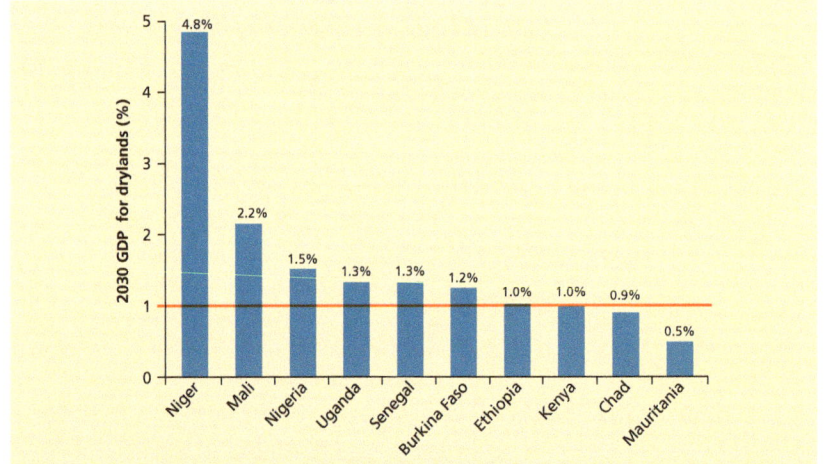

*Source:* Calculation based on the approach discussed in the Appendix.

*Note:* The chart shows the cost (expressed as a percentage of 2030 GDP for drylands, assumed proportional to the share of the population living in drylands) of bringing all drought-affected people to the international poverty line through cash transfers. The cost is calculated taking into account the country-specific depth of poverty, as proxied by 2010 poverty gap index obtained from the World Bank PovCalnet database. Figures for 2030 GDP are based on the average growth scenario as defined in Carfagna, Cervigni, and Fallavier (2016). The reference line (1 percent of GDP) indicates the consensus value in the social protection literature on the resources governments should be willing to spend in social safety nets.

of normal rainfall and rising in drought years when the number of drought-affected people surges. Second, the cost estimates implicitly assume that social safety net support can be perfectly targeted to drought-affected households; in practice it is very difficult to ensure that safety net support reaches *all* drought-affected households and *only* those households, and in the presence of leakages, overall financing needs would be considerably higher.

In conclusion, it is safe to assume that for most dryland countries, relying on social protection instruments to protect vulnerable populations against the effects of drought shocks will likely be beyond their fiscal means and institutional capacity.

## Estimating the potential for enhancing resilience

The umbrella model results show that it would be prohibitively expensive for governments in dryland countries to rely on social safety nets to protect vulnerable households from the adverse effects of droughts and other shocks. In that context, policy makers will want to know to what extent the coming challenge can be mitigated by making current livelihood strategies more resilient.

To help determine the answer, a set of best-bet interventions was selected from among the many resilience-enhancing interventions reviewed in previous chapters (table 12.1), and the umbrella model was used to assess the extent to which these interventions would be able to reduce vulnerability among populations living in drylands by improving the productivity and sustainability of current livelihood strategies.

Because of technical limitations in the umbrella model, which does not have the capacity to capture complex interactions that occur when multiple interventions are implemented simultaneously, only the livestock-related interventions were considered in hyper-arid and arid zones, and only the crop farming-related interventions were considered in semi-arid and dry subhumid zones. This approach ignores the significant scope for implementing livestock-related interventions in agropastoral systems found in semi-arid and dry subhumid zones. For this reason, while the modeling results indicate the order of magnitude of the likely resilience benefits of the different interventions, they represent conservative lower bound estimates of the full potential.

The results of the modeling exercise suggest that the best-bet interventions, by improving the productivity of livestock and crop farming systems in the drylands, could considerably slow the projected increase in the number of drought-affected people (figure 12.4). Without the interventions, by 2030 the number of drought-affected people is projected to increase by 60 percent compared to 2010. With the interventions, the number of drought-affected people is projected to increase by

**Table 12.1** Coverage of resilience interventions in umbrella model

| Livelihood | Intervention | Hyper-arid, arid | Semi-arid, dry subhumid |
|---|---|:---:|:---:|
| Livestock-based | Improved animal health services | ☑ | |
| | Early offtake of young male animals | ☑ | |
| Farming-based and mixed | Drought-tolerant germplasm | | ☑ |
| | Heat-tolerant germplasm | | ☑ |
| | Soil fertility management | | ☑ |
| | Agroforestry/FMNR | | ☑ |
| | Heat-tolerant germplasm and FMNR | | ☑ |
| | Drought-tolerant germplasm and soil fertility management | | ☑ |
| | Drought-tolerant and heat-tolerant germplasm | | ☑ |
| | Irrigation | | ☑ |

*Source:* Calculation based on the approach discussed in the Appendix.
*Note:* FMNR = farmer-managed natural regeneration.

**Figure 12.4** Contribution of technical interventions to resilience in 2030 (2010=100%)

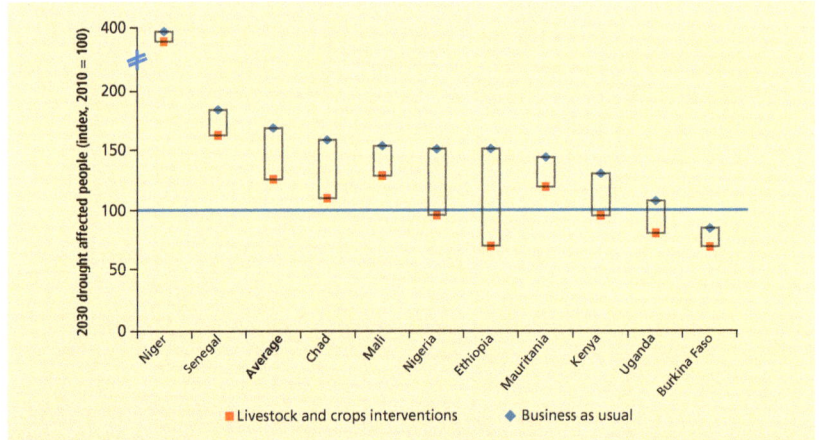

*Source:* Calculation based on the approach discussed in the Appendix.

*Note:* The vertical axis in the figure has been trimmed to accommodate Niger, an extreme outlier. The number of drought-affected farmers was estimated using the African Risk Capacity (ARC) model, based on the yields obtained for a set of reference staple crops (maize, sorghum, millet) grown with and without the interventions, and evaluating the number of years (over a 20-year simulated time series reflecting historical climate) in which yields fall below a certain threshold. The number of drought-affected herders was estimated based on the number of households that are able to sustain—for a given amount of biomass determined by historical climate patterns—a minimum herd size; herders lacking the minimum herd size were assumed to take up crop farming (which may or may not have made them resilient).

only 27 percent (an improvement of 43 percentage points). In some countries, notably Ethiopia and to a lesser extent Kenya and Nigeria, by 2030 adoption of improved management of livestock and crop farming systems could reduce the absolute number of drought-affected people relative to the 2010 baseline. In other countries, particularly Niger but also Senegal and Mauritania, the best-bet interventions would have a more modest impact, and the number of drought-affected people in 2030 would still be considerably larger than in 2010.

In pastoral areas, where only livestock-related interventions were considered (specifically, improved animal health services and early offtake of young male animals), the most important benefit is to slow the exit of the poorest herders who otherwise would be forced to abandon pastoralism and take up other livelihood activities (mainly crop farming). By increasing livestock productivity, the livestock-related interventions reduce the minimum number of TLU needed to generate the amount of income required by livestock-dependent households to remain above the poverty line. In this way, the livestock-related best-bet interventions are projected to reduce the number of exits from pastoralism by 6 percent on average. The effect is much higher in some countries: Kenya (13 percent fewer exits), Burkina Faso (14 percent fewer exits), Mali (16 percent fewer exits), and Ethiopia (19 percent fewer exits) (figure 12.5).

**Figure 12.5** Reduction in exits from pastoralism due to technical interventions, 2030 (%)

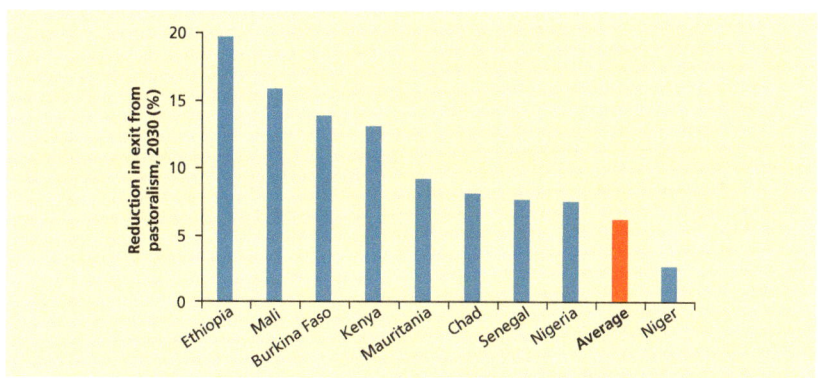

*Source:* De Haan 2016.

With respect to crop farming interventions, the biggest impact on reducing the number of drought-affected people is projected to come from improvements in soil fertility management, followed by irrigation development, adoption of drought-resistant varieties, and uptake of tree-based systems. The benefits of the different crop farming interventions vary considerably by location, and the mix of optimal interventions is quite variable across countries, pointing to the importance of carrying out location-specific assessments and tailoring interventions to meet local circumstances (figure 12.6).

**Figure 12.6** Relative contributions of technical interventions in the reduction of vulnerability, by country, 2030 (%)

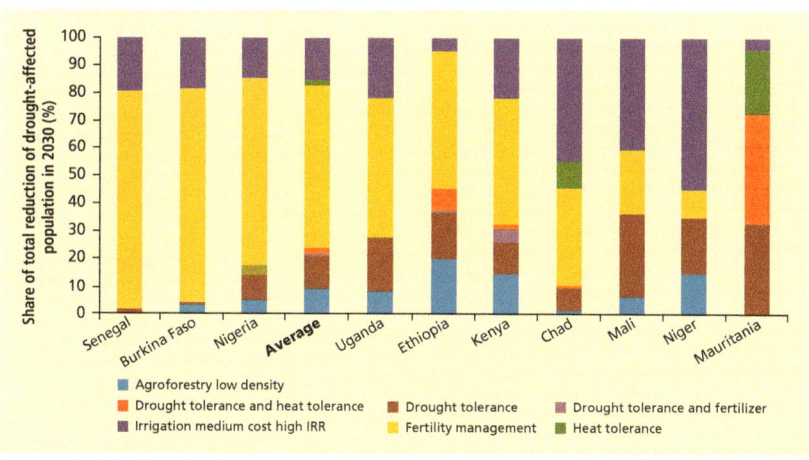

*Source:* Calculation based on the approach discussed in the Appendix.
*Note:* IRR = internal rate of return.

Not surprisingly, the mix of optimal crop farming interventions varies by arid-ity zone (figure 12.7). In the drier parts of the semi-arid zone (Aridity Index 0.2–0.35), irrigation development is likely to have the largest impact, followed by adoption of soil fertility management practices and drought-tolerant varieties. In the wetter parts of the semi-arid zone and the dry subhumid zone (Aridity Index 0.36–0.65), adoption of fertility management practices is likely to have the biggest impact by far. Adoption of tree-based systems/FMNR is likely to have a larger impact in the dry subhumid zone compared to more arid zones.

One important positive message emerging from the umbrella modeling work is that when accurately targeted, the best-bet crop farming interventions have considerable potential to reduce the impacts of droughts. Accurate targeting was ensured in the umbrella modeling work by restricting implementation of the interventions only to locations in which their adoption was determined to be profitable (that is, simulated yield gains remain positive after yields have been adjusted downward to reflect the cost of adopting the technology).

A second important message emerging from the umbrella modeling work—admittedly less positive—is that it is critically important to accurately target the crop farming interventions to the locations where they will have maximum impact. The importance of accurate targeting was determined by rerunning the umbrella model and allowing the interventions to be implemented in all locations regardless of profitability. The results of this second set of model runs (summa-rized in figure 12.8) make clear that the cost of inaccurate targeting can be high.

**Figure 12.7** Relative contributions of technical interventions in reducing vulnerability, by aridity zone, 2030 (%)

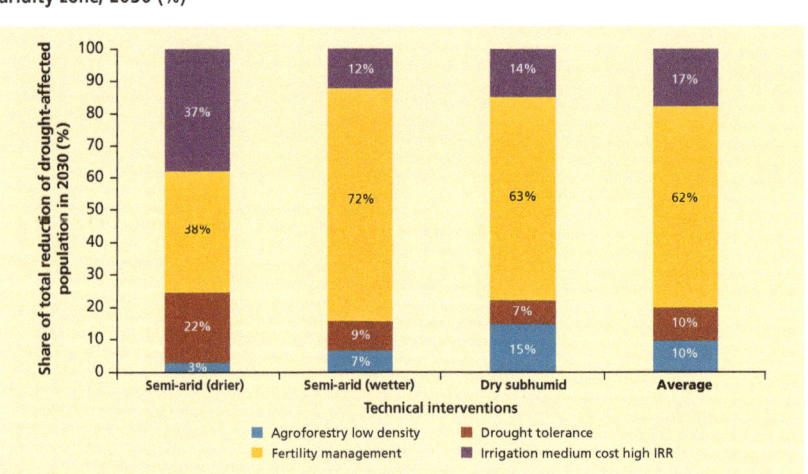

Source: Calculation based on the approach discussed in the Appendix.

Note: IRR = internal rate of return

**Figure 12.8** Importance of targeting technical interventions (BAU=100)

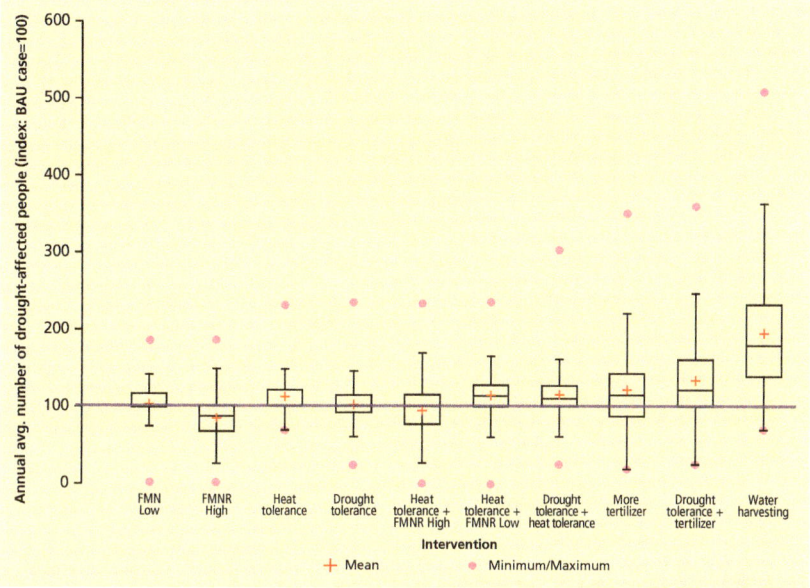

*Source:* Calculation based on the approach discussed in the Appendix.

*Note:* The Y-axis values indicate for each technology the distribution across locations of the number of drought-affected people, expressed as a percentage of the business as usual (BAU) case. Values above 100 indicate poorer performance than BAU (more drought-affected people, suggesting that in the corresponding areas it is thus better not to adopt the technology); values below 100 indicate better performance than BAU (fewer drought-affected people, suggesting that in those areas it makes sense to adopt the technology). The larger the portion of the box above the 100 line, the larger the chance that the corresponding technology will result in an increase of the average annual number of drought-affected people. FMNR = farmer-managed natural regeneration.

In figure 12.8 the vertical axis represents the projected number of drought-affected people living in drylands in 2030, expressed as a percentage of the BAU scenario. Values above 100 represent an increase in the number of drought-affected people relative to the BAU scenario, and values below 100 represent a decrease in the number of drought-affected people relative to the BAU scenario. As shown in figure 12.8, many of the best-bet crop farming interventions are expected to reduce the number of drought-affected people only in selected locations. In many other locations the cost of adopting the crop farming intervention does not justify the expected benefits, resulting in a net loss in income and leaving adopting households more likely to be adversely affected by droughts. This means that careful assessments will need to be made to ensure that the best-bet crop farming interventions are promoted only in locations in which they will actually deliver benefits (that is, increasing resilience to drought shocks).

## Evaluating the costs of resilience interventions

Dryland development policies must take into account not only the extent to which interventions can reduce vulnerability and increase resilience, but also the cost of implementing those interventions. Since the best-bet interventions considered in this book are already available "on the shelf" and are ready for implementation, research and development costs are sunk costs and can safely be ignored. Additional costs that need to be considered include:

1. Private costs associated with technology adoption (e.g., the costs incurred by herders and farmers when purchasing inputs and/or hiring additional labor)
2. Public costs associated with technology transfer (e.g., the costs of extension campaigns and farmer training)
3. Miscellaneous overhead costs

Because technology transfer costs vary considerably depending on the accuracy of targeting, these costs were estimated for three scenarios:

1. Zero targeting: All technologies are promoted in all polygons having non-zero cropping area.
2. Intermediate targeting: All technologies are promoted only in polygons having non-zero cropping area and in which farm-level benefits exceed technology transfer costs (see the Appendix for details).
3. Perfect targeting: Among the technologies having positive farm-level benefits, the only technology that is promoted is the one with the greatest impact on resilience, that is, the one producing the largest reduction in the number of drought-affected people.

Depending on accuracy of targeting, the average annual cost across the entire sample of dryland countries ranges from US$0.14–1.31 billion (table 12.2).

Costs on this order of magnitude compare favorably with current levels of development assistance provided in dryland countries.

## Do investments in resilience pay off?

How cost-effective are these best-bet interventions compared to alternative strategies for reducing vulnerability and increasing resilience in the drylands? To answer this question, a simple benefit/cost (B/C) assessment was carried out in which the cost of the interventions was compared to the benefits, measured in terms of the savings that would be achieved in the amount of safety net cash transfers that would be needed to bring all drought-affected people to the poverty line. The B/C analysis assumed the following:

**Table 12.2** Estimated annual costs of resilience interventions (US$ billions)

| Cost Item | No targeting | Partial targeting | Perfect targeting | Other |
|---|---|---|---|---|
| Private – livestock and crops | 1.09 | 0.36 | 0.12 | |
| Private – irrigation | | | | 2.18 |
| Public | 0.21 | 0.06 | 0.02 | |
| Total | 1.31 | 0.43 | 0.14 | |

*Source:* Calculation based on the approach discussed in the Appendix.

*Note:* Irrigation costs are reported separately because the targeting of irrigation investments is "built-in" to the analysis, which assumes that irrigation development occurs only in locations where the investment is expected to generate an internal rate of return (IRR) of 12 percent or more (see Chapter 5 for details).

- In the no-intervention scenario, the number of drought-affected people would increase in linear fashion over 15 years, as would the corresponding increase in cash transfers needed to lift these people out of poverty; meanwhile, no expenditure would be made in the best-bet interventions (see figure 12.2).

- In the intervention scenario, the cash transfers needed to lift drought-affected people out of poverty would increase more slowly, commensurate with the slower increase in the number of drought-affected people; meanwhile, public investment in the best-bet interventions would increase in linear fashion, with the cumulative expenditure over the 15 years equaling the sum of the annual averages. The total public investment was calculated as the sum of the technology transfer cost and the overhead cost, plus 25 percent of the private cost (representing subsidies needed to encourage adoption).

- In the intervention scenario, intermediate targeting was assumed; this implies that public agencies will be able to prescreen investments and avoid promoting technologies that are poorly suited to local agro-climatic circumstances, but they will lack the ability to identify and exclusively promote the best-performing technology in any given location.

- In the intervention scenario, a cost escalation factor was used to carry out sensitivity analysis in recognition that technology transfer costs have been crudely estimated and could change significantly in future; the cost escalation factor varies from 1 (no cost escalation) to 4 (four-fold increase in technology transfer costs).

- In both scenarios, the discount rate was set at 10 percent.

The results of the B/C assessment suggest that the benefits—expressed in terms of reduced cash transfers needed to support drought-affected people—far exceed the costs of implementing the best-bet interventions (figure 12.9). In most countries (the only exceptions are Mauritania and Niger), the results are robust under a wide range of cost assumptions: even if costs are increased four-fold, the B/C ratio remains well above 1.

**Figure 12.9** Benefit/cost ratios of resilience interventions (log)

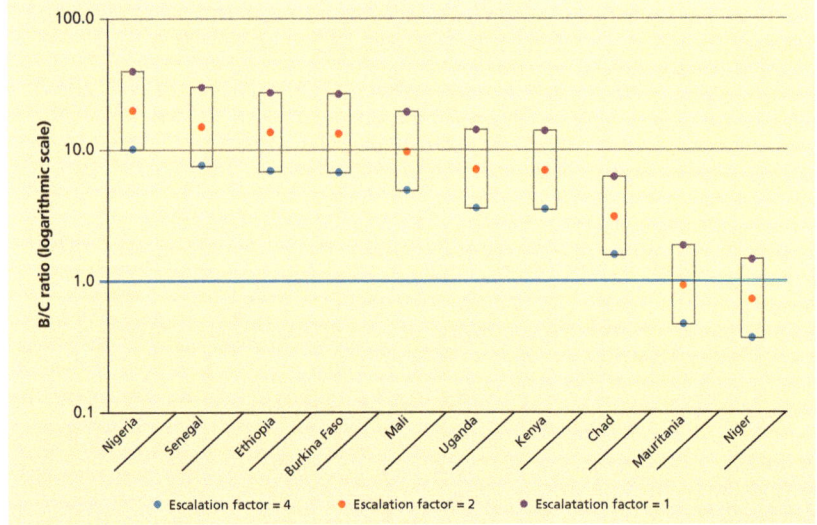

*Source:* Calculation based on the approach discussed in the Appendix.

*Note:* Benefit-cost (B/C) ratios above 1 (the horizontal line on the chart) indicate that the benefits of resilience interventions exceed the costs.

## Are investments in existing livelihoods sufficient to ensure resilience?

Will the best-bet interventions identified in this book be able to solve in a lasting manner the challenge of resilience in drylands?

Before addressing this question, it is important to agree on what might be considered an acceptable outcome. The policy objective in drylands cannot be to eliminate completely the need to provide support to people who have been adversely affected by droughts: drylands will always be subject to droughts, and for the foreseeable future, significant numbers of people will be exposed to droughts, sensitive to their effects, and unable to cope in their wake.

In that context, a reasonable policy objective would be to ensure that support is adequate (in the sense of covering those in need) and manageable (in the sense of remaining within the country's long-term fiscal capacity). Using again the metric of the cost of providing cash transfer support to drought-affected people, it is useful to see how implementation of the best-bet interventions compares to the BAU scenario (figure 12.10).

In considering the potential of the best-bet interventions to reduce vulnerability and increase resilience among populations living in the drylands, it is worth noting that the interventions will have two types of effects—a direct effect

**Figure 12.10** Cost as share of GDP of supporting drought-affected people in drylands (with and without interventions) (%)

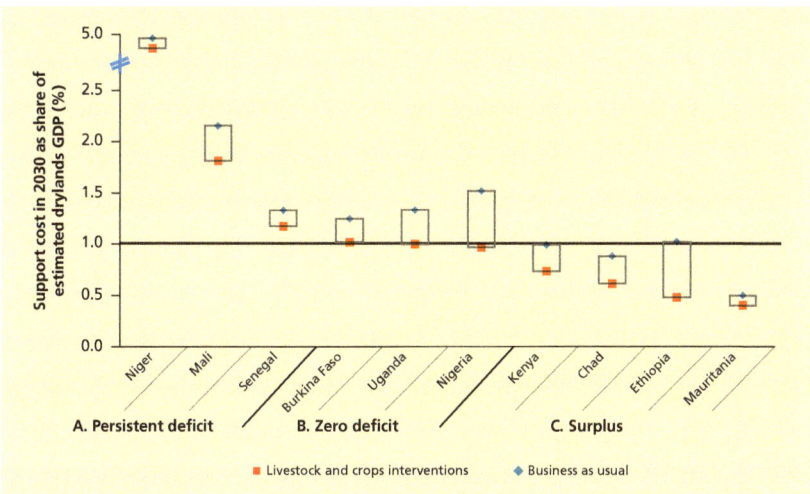

*Source:* Calculation based on the approach discussed in the Appendix.

*Note:* The vertical axis has been trimmed to avoid the distorting effect of the outlier (Niger). The chart indicates the cost (expressed as a percentage of 2030 GDP for drylands, assumed proportional to the share of the population living in drylands) of bringing all drought-affected people to the international poverty line, without interventions (business as usual), and with interventions. The cost is calculated taking into account the country-specific depth of poverty, as proxied by the 2010 poverty gap index obtained from the World Bank PovCalnet database. Figures for 2030 GDP are based on the average growth scenario as defined in the umbrella model (see Appendix for details). The reference line (1 percent of GDP) indicates the consensus value in the social protection literature on the resources governments should be willing to spend in social safety nets.

and an indirect effect. Investments in livestock and crop farming systems will directly reduce the number of drought-affected people by improving the productivity and sustainability of current livelihood strategies. In addition, these investments will indirectly contribute to improved resilience in the drylands by freeing up public resources that would otherwise have to be used for emergency responses; these resources can instead be redirected to other programs to strengthen the resilience of vulnerable segments of the population.

The overall impact of the resilience-enhancing interventions varies considerably across the different countries. Three main outcomes can be distinguished, as follows.

- In Niger, Mali, and Senegal the resilience-enhancing interventions reduce the cost of having to rely on safety nets to support the drought-affected population, but the cost of safety nets remains well above the 1 percent of GDP mark so that a wide resilience deficit persists.

- In Burkina Faso, Uganda, and Nigeria the resilience-enhancing interventions reduce the cost of having to rely on safety nets to support the drought-affected population to approximately the 1 percent of GDP mark, so the drought-affected population is covered in the short run, but very little financing is left from the 1 percent of GDP for investment in programs that can build resilience over the longer term and reduce the need for future safety net expenditures.

- In Kenya, Chad, Ethiopia, and Mauritania the resilience-enhancing interventions reduce the cost of having to rely on safety nets to support the drought-affected population to well below 1 percent of GDP, meaning that drought-affected populations can be covered in the short run even as resources are freed for investment in programs that can build resilience over the longer term and reduce the need for future safety net expenditures.

These findings have implications for the mix of policies and instruments that each group of countries will want to consider in seeking to ensure that dryland populations remain resilient in the face of droughts and other shocks (see also table 12.3).

**Niger, Mali, and Senegal (referred to here as Group A).** The resilience-enhancing interventions discussed in this book, albeit cost-effective in terms of reducing the number of drought-affected people, will likely be insufficient to bring drought management down to a fiscally sustainable footing. An important priority for policies in these countries will be to identify additional interventions to strengthen existing livelihoods beyond those that the limited scope of this book permitted. But perhaps more importantly, public policies could actively explore opportunities to develop alternative livelihoods, both inside and outside of drylands (more on this in the next chapter).

**Burkina Faso, Uganda, and Nigeria (referred to here as Group B).** The resilience-enhancing interventions discussed in this book, while capable of reducing the numbers of drought-affected people, will leave significant numbers of drought-affected people reliant on safety net support. If the governments in these countries are willing to allocate 1 percent of GDP on average to safety net support, it should be possible to provide assistance to all drought-affected people. These resources will have to be managed carefully, however, because the actual financing needs will fluctuate significantly from year to year. An important priority for these countries, in addition to exploring a wider range of options for strengthening existing livelihoods, will be to develop mechanisms for mobilizing contingent finance (e.g., sovereign insurance) when social protection needs exceed their ability to finance it.

**Kenya, Chad, Ethiopia, and Mauritania (referred to here as Group C).** The resilience-enhancing interventions discussed in this book will be able to significantly reduce the number of drought-affected people, leaving a relatively small number reliant on safety net support. If the governments in these countries are willing to allocate 1 percent of GDP on average to social protection programs, a "dividend" will remain that could be invested in activities designed to improve their livelihood strategies and achieve permanent income gains. Key priorities for these countries are to scale up investments in resilience-enhancing interventions, as well as to define strategies for sustainably reinvesting the additional dividends that these interventions will generate.

**Table 12.3** Policy priorities to ensure resilience, selected dryland countries

| Countries | Priority interventions | Safety net coverage achieved with 1% of GDP | Fiscal dividend remains | Importance of promoting alternative livelihoods |
|---|---|---|---|---|
| **Group A**<br>Mali<br>Niger<br>Senegal | **Semi-arid zones**<br>• Livestock health<br>• Livestock early offtake<br>• Drought tolerance<br>• Irrigation<br>• Tree-based systems<br>**Dry subhumid zones**<br>• Fertilizer<br>• Irrigation<br>• Tree-based systems | Drought-affected people not fully covered | No | High |
| **Group B**<br>Burkina Faso<br>Nigeria<br>Uganda | **Semi-arid zones**<br>• Livestock health<br>• Livestock early offtake<br>• Drought tolerance<br>• Irrigation<br>• Tree-based systems<br>**Dry subhumid zones**<br>• Fertilizer<br>• Irrigation<br>• Tree-based systems | Drought-affected people just covered | No | Medium |
| **Group C**<br>Chad<br>Ethiopia<br>Kenya<br>Mauritania | **Semi-arid zones**<br>• Livestock health<br>• Livestock early offtake<br>• Drought tolerance<br>• Irrigation<br>• Tree-based systems<br>**Dry subhumid zones**<br>• Fertilizer<br>• Irrigation<br>• Tree-based systems | Drought-affected people fully covered | Yes | Low |

# References

Carfagna, F., R. Cervigni, and P. Fallavier, eds. 2016 (forthcoming). *Mitigating Drought Impacts in Drylands: Quantifying the Potential for Strengthening Crop- and Livestock-Based Livelihoods.* World Bank Studies. Washington, DC: World Bank.

De Haan, C., ed. 2016. *Improved Crop Productivity for Africa's Drylands.* World Bank Studies. Washington, DC: World Bank.

Chapter **13**

# The Road Ahead: Toward a Shared Dryland Development Agenda

*Michael Morris, Raffaello Cervigni, Karen Brooks*

## Scope of the dryland development challenge

The chronic vulnerability of people living in drylands stands at the forefront of Africa's development challenge. Drylands make up 43 percent of the total land area in Sub-Saharan Africa, account for 75 percent of the area used for agriculture, and are home to about 50 percent of the region's total population. Poverty is heavily concentrated in drylands: about 75 percent of Africans living on less than US$1.25 per day live in countries in which at least one-quarter of the population lives in dryland zones.

In the drylands today frequent and severe shocks, both natural and human-induced, limit the livelihood opportunities available to millions of households and undermine efforts to eradicate poverty. In the absence of robust social protection systems and rapidly scalable safety nets, these shocks regularly cause large drains on government budgets and consume a significant portion of the region's international development assistance. They have also contributed to a pronounced development gap: the people living in drylands are less wealthy than those living outside of drylands, less healthy, less educated, and less secure.

Over the next two decades, if current trends continue, dryland regions of Africa will experience strong population growth. By 2030 the population in drylands is expected to grow by 58–74 percent (depending on the fertility scenario), putting increased pressure on a resource base that is already stretched. Over the same period, climate change could result in an expansion of the area classified as drylands (by as much as 20 percent under some scenarios), bringing more people into an ever more challenging environment. Higher population density in the drylands will put additional pressure on a fragile natural resource base, pushing it in some cases beyond its natural regenerative capacity. As competition for resources intensifies, conflicts over land, water, and feed resources are likely to multiply.

These trends lead to an inevitable conclusion: Business as usual is not an option. Thus there is a need for African governments and the broader development community to bring fresh thinking and new ideas to a longstanding problem that continues to defy conventional development solutions. This book has attempted to make a contribution toward that objective.

## Demographic trends: Challenges and opportunities

Most developing regions of the world have experienced gradually declining rates of population growth. In Sub-Saharan Africa, however, the population is still growing rapidly, because the region has not yet embarked on its "demographic transition" from high to low birth and death rates. In the countries of the Sahel and the Horn of Africa, demographic growth rates range from 2.5 percent to nearly 4 percent per year, and if these high growth rates persist, the population will continue to increase rapidly. The high growth rates in these countries are occurring because infant and child mortality rates are decreasing rapidly, while fertility rates are falling much more slowly.

As a result of the generally high population growth rates, the age structure of the populations of the countries of the Sahel and the Horn of Africa will remain predominately young for the foreseeable future. The number of youth—those younger than 20—will double by 2050, and youth dependency ratios (defined as the ratio of youth to people of working age) will remain among the highest in the world. The demographic dividend that could be gained from a larger workforce, when relatively more working adults support relatively fewer dependents, appears to be decades away for the majority of the countries of the Sahel and the Horn of Africa, although countries are looking into ways to address the challenge (box 13.1).

At the same time, it is important to recognize that in addition to creating challenges, demographic trends may also bring opportunities. These include a larger market size for commerce and trade, new opportunities for increased economic specialization, and unprecedented possibilities for enhanced value addition. Similarly, increased population density in the drylands will reduce the cost of providing essential public services, such as education, health care, water and sanitation, communications, and security. For these reasons population growth in the drylands could prove vital in overcoming the traditional problem that has contributed to the underdevelopment of the drylands—namely, that the sparse population distributed over vast areas has made markets thin and costly, discouraging both public and private investment in goods and services.

**BOX 13.1**

### Initiatives to address the challenge of population growth

Governments in Sub-Saharan Africa and their development partners can, and are, taking actions to rein in population growth. A good example is the Ouagadougou Partnership, which was founded in February 2011 following the Regional Conference on Population, Development, and Family Planning and which is committed to two principles: (1) increasing donor coordination so as to strengthen support for partnership countries, and (2) collaborating at national and regional levels to address the unmet need for family planning. With support from a number of development partners, the nine countries of the Partnership are working to improve female education, inform populations about the benefits of smaller family size, improve access to contraceptives, and raise the legal age of marriage. If realized, these changes will reduce the strain on the natural resource base, allow for more educational investments in young people, and create the possibility of a demographic dividend in the long run.

Initiatives to slow the rate of population growth can help relieve the growing pressure on the resource base in the drylands, tempering the projected increases in the numbers of vulnerable people and reducing the challenge of building resilience among those who will continue to rely on traditional livelihoods activities. It is important to recognize, however, that while slowing the rate of population growth can reduce the rate at which the number of vulnerable households increases in future, it will not be able to reduce the absolute number of vulnerable households, which will continue to rise. Initiatives to slow population growth therefore represent an important element in the effort to reduce vulnerability and build resilience in the drylands, because they can buy time to implement policies and programs that build resilience, but they are not a silver bullet that will be able to solve the problem on their own.

Seizing the emerging opportunities will be possible only to the extent that higher population density does not lead to increased competition for natural resources, especially land, water, and biomass. Increased competition would likely lead to erosion of the resource base and eventually give rise to conflict. For this reason, as population growth outstrips the ability of current livelihood strategies to provide adequate incomes for all, public policy will have to focus on the creation of new livelihoods, less reliant on natural capital, and more on human and physical capital.

## Demographic trends will require new livelihood strategies

Evidence presented in this book suggests that the predominant livelihood strategies in the drylands will have to change. Higher population density is not consistent with continued widespread reliance on traditional dryland livelihood strategies such as livestock-keeping and agriculture, which are based on harvesting of ecosystem services and are very heavily reliant on natural capital. The natural resource base will not be able to support denser populations without degradation and competition for resources leading to conflict.

As population growth outstrips the carrying capacity of the natural resource base on which most current livelihoods depend, livelihood strategies will have to shift to activities more reliant on human and physical capital, and that complement use of natural resources with other inputs. This shift implies a gradual transition, not an abrupt wholesale conversion of large numbers of people from one set of activities to a different set of activities. Traditional livelihood strategies will need to evolve by adding human and physical capital to make use of natural resources more productive and sustainable.

As part of the larger transformation, significant numbers of people will have to exit from agricultural and natural resource-based livelihoods to seek employment in other sectors. Among other things, this means that the solution to dryland problems will come to a large extent from outside the drylands.

## Impacts of climate change

Adding to the uncertainty about the future prospects for those living in drylands is the prospect of climate change. While difficult to predict with certainty, the preponderance of evidence suggests that climate change is likely to have a significant impact on Sub-Saharan Africa. The effects are likely to include shifts in the distribution of drylands, expansion in the size of drylands, and increases in the frequency and severity of extreme weather events experienced within drylands. Depending on the rate at which these projected impacts of climate change manifest themselves, over time the number of vulnerable people living in dryland regions of Africa is likely to increase even further.

Climate change will exacerbate the need for a shift in livelihood strategies, but the shift would have been needed anyway. Climate change will mainly alter the angle of the trajectory and accelerate the rate of needed changes.

## Public policy priorities: Short term

The coming transformation of the drylands will not happen overnight. In the short run, possibilities for migration are severely restricted, because few high-quality jobs are being created outside the rural sector. The implication is that for the foreseeable future, many people will have to remain in the drylands, relying primarily on agricultural and natural resource-based livelihoods.

The fact that current livelihood strategies will remain vital for the foreseeable future has important implications for policy priorities in the short to medium term. Information and analysis presented in this book show clearly that opportunities exist to make agricultural and natural resource-based livelihoods more productive, more stable, and more sustainable. Governments and their development partners must act quickly to make certain that these opportunities are fully exploited.

This book has: identified a series of best-bet interventions with the potential to improve the productivity and sustainability of current livelihood strategies; estimated the extent to which these interventions could reduce vulnerability and strengthen resilience among people living in drylands; and calculated the approximate cost of fully implementing these interventions. The most promising of these interventions, along with the key policy recommendations needed to ensure their successful implementation, can be summarized as follows.

## Livestock

Livestock systems in the drylands can be made more productive and more profitable, but ensuring the resilience of all pastoralists and agro-pastoralists will require the addition of new income sources.

*Key recommendations:*

1. Increase production of meat, milk, and hides in drylands by developing sustainable delivery systems for animal health, promoting increased market integration, and exploiting complementarities between drylands and higher rainfall areas.

2. Enhance the mobility of herds by ensuring adequate and equitable year-round access to grazing and water and by improving security in pastoral zones.

3. Develop Livestock Early Warning Systems (LEWSs) and early response systems to reduce the adverse impacts of shocks.

4. Identify additional and alternative livelihood strategies, because feed and animal resources will be insufficient in the drylands to enable the minimum level of herd ownership needed to provide adequate income, food security, and asset-building opportunities for all livestock-keeping households.

## Rainfed agriculture

Improved crop production technologies can deliver sizeable resilience benefits by boosting productivity in rainfed agriculture, but only if barriers to adoption can be overcome.

*Key recommendations:*

1. Accelerate the rate of varietal turnover.
2. Increase availability of hybrids.
3. Improve soil fertility management.
4. Improve agricultural water management.

## Irrigated agriculture

Irrigation can provide an important buffer against drought in dryland areas, but only for a relatively small share of the population. Irrigation development is technically feasible and financially viable on 5–10 million hectares in the drylands (the number varies depending on assumptions made about capital investment costs and the minimum required level of financial returns). Prospects are brighter for small-scale irrigation, due in large part to the more modest investment costs.

*Key recommendations:*

1. Give a more prominent role to agricultural water management in development planning.
2. Promote development of small-scale irrigation, especially in areas where cash crops are produced and farmers have access to markets where they can sell their production.
3. Triple the area developed for large-scale irrigation, rehabilitating existing capacity that is currently underutilized and adding a further 10 million hectares of irrigation development to the current 5 million hectares.

## Tree-based systems

Tree-based systems include both those based on farmer management of naturally occurring species (generally more appropriate in drier zones), as well as those involving deliberate planting of economically useful species (generally more appropriate in more humid zones).

*Key recommendations:*

1. Promote farmer-managed natural regeneration (FMNR) to establish a range of beneficial trees throughout the drylands.
2. Invest in tree germplasm multiplication and promote planting of location-appropriate high-value species especially in dry subhumid areas.
3. Develop value-added opportunities for the many valuable tree products produced in the drylands.

## Social protection

Even under optimistic assumptions about the spread of resilience-enhancing interventions such as those described above, a significant share of the population living in drylands will remain vulnerable to shocks. Since it is unlikely that the entire dryland population can be made resilient in the face of every type of shock, social protection programs including safety nets will be needed to support the most disadvantaged households and those affected by disasters.

*Key recommendations:*

1. Establish and gradually expand the coverage of national adaptive safety net programs that promote resilience of the poorest households.
2. Use social protection programs to build capacity of vulnerable households to climb out of poverty, but maintain the ability to provide humanitarian assistance in the short run.
3. Respond to emergencies by scaling up existing programs, rather than relying on appeals for humanitarian assistance.
4. Tailor social protection programs to address the unique circumstances of dryland populations.

## Public policy priorities: Medium to long term

Short- and medium-term measures designed to improve the productivity and stability of current agricultural and natural resource-based livelihood strategies and to ensure their sustainability will have to be complemented with long-term measures to facilitate the transformation. Two broad sets of interventions will be needed as follows.

Public policy will need to encourage investment in human capital:

- Education and vocational training
- Health and nutrition
- Fertility management

Public policy will also need to encourage investment in physical capital:

- Transport infrastructure
- Communications
- Housing (urban focus)

## Roles and responsibilities of non-state actors

Governments will have to play a leading role in managing the coming transformation, but governments will need help. Changing the trajectory will require cooperative efforts on the part of development partners, the private sector, and civil society.

**Development partners** can contribute through investments designed to facilitate sustainable intensification of dryland livelihoods, as well as through implementation of supportive policies relating to improvement of health care services/fertility control, education, migration, and foreign investment, among others.

**The private sector** can contribute primarily through investments designed to facilitate sustainable intensification of dryland livelihoods, especially by creating jobs in non-dryland areas to absorb exits from drylands.

**Civil society leaders** can play an important role in encouraging behavior change and attitudinal adjustment, for example by building support for girls' schooling and secular education generally, facilitating changes in traditional land use patterns of pastoral peoples, and mediating conflicts at the local level over competition for natural resources.

## Final thoughts

Today in the Sahel and the Horn of Africa frequent and severe shocks, both natural and human-induced, limit the livelihood opportunities for millions of poor and vulnerable households, undermining efforts to eradicate poverty and break the recurring cycle of humanitarian crises. This book has focused on quantifying the dimensions of the challenge likely to confront African governments in the coming decades, as well as assessing the scope for public policy interventions to reduce vulnerability and increase resilience of dryland populations by improving the productivity and ensuring the sustainability of current livelihood strategies. The impact of these interventions must be understood within a wider context of the long-term transformational change that drylands are already experiencing.

Interventions such as those discussed in this book will be able to reduce the vulnerability of many people living in drylands, but they will not be sufficient. Additional measures that generate employment opportunities outside of agriculture and equip rural populations with the skills to take advantage of those opportunities will be needed as well. Over time these additional measures will provide relief by helping to accelerate the inevitable structural transformation of dryland economies.

Successful management of the ongoing structural transformation will allow socially desirable outcomes to be realized, but the challenges are very large. Without constructive engagement of public officials, development partners, and civil society at many levels, adverse outcomes are possible and even likely.

The stakes are high. Opportunities are emerging to build vibrant societies incorporating both the traditional and the new, but if these opportunities are missed, there is a very real possibility that the people living in the drylands will be condemned to many more decades of poverty, immiseration, and conflict.

# Technical Note on the Drought Impacts Model*

*Raffaello Cervigni, Michael Morris, Federica Carfagna, Joanna Syroka, Balthazar de Brouwer, Elke Verbeeten, Jawoo Koo, Pierre Fallavier, Hua Xie, Weston Anderson, Nikos Perez, Claudia Ringler, Liang You*

How many people live in dryland zones of Sub-Saharan Africa, and what are their livelihood strategies? How many of these people are vulnerable to droughts and other shocks, and of those who are vulnerable, how many are actually affected in an average year? How are the numbers of vulnerable and drought-affected people living in drylands likely to evolve as the population increases and national economies grow and transform? To what extent can the impacts of drought be mitigated through policy interventions that improve the productivity and sustainability of livelihood strategies or provide protection in the form of safety nets? And how much would these policy interventions cost?

These questions are hard to answer, for two main reasons. First, because national statistical reporting services in many dryland countries are weak, detailed information is not always available either about the people who currently live in the drylands or about their livelihood activities. Second, because events in the drylands are influenced by a complex set of agro-climatic, demographic, economic, and political drivers, projecting future trends is technically difficult.

Despite these challenges, the team that carried out the Africa Drylands Study made an effort to quantify the scope of the challenge facing policy makers, with the objective of providing insight into the likely impacts and fiscal costs of alternative resilience-enhancing interventions. Answers to the above questions were generated with the help of a diverse set of modeling tools.

The modeling effort proceeded in four stages:

1. Estimation of the 2010 baseline population [umbrella model]
2. Projection of population growth to 2030 [umbrella model]

---

*Technical details of the modeling approach are described more extensively in Carfagna, Cervigni, and Fallavier (2016). Unless otherwise noted, all figures and tables are based on the umbrella model.

3. Modeling of likely effects of resilience interventions targeting:
    a) Livestock systems [livestock model]
    b) Rainfed cropping systems [cropping model]
    c) Irrigation systems [irrigation development model]
4. Consolidation of results [umbrella model]

This appendix provides details of the modeling tools, describes the data used for the simulations, explains key assumptions underlying the analysis, and discusses strengths and weaknesses of the approach.

## Geographical coverage

Before considering the modeling tools, it is useful to review the geographical coverage of analysis.

### Definition of drylands
For reasons of simplicity and for consistency with widespread common practice, "drylands" are defined on the basis of the Aridity Index (AI). Under this approach, which has been endorsed by the 195 parties to the United Nations Convention to Combat Desertification (UNCCD) and which also is being used by the United Nations Food and Agriculture Organization (FAO), drylands are defined as regions having an AI of 0.65 or less. Drylands are furthermore subdivided into four zones: hyper-arid, arid, semi-arid, and dry subhumid. In some of the analysis (e.g., assessment of the effectiveness of crop farming interventions), the semi-arid zone is additionally divided into a "dry semi-arid zone" and a "wet semi-arid zone." The Aridity Index ranges used to define these zones appear in table A.1.

### Country coverage
Because the various analyses required different types of information, the coverage varied depending on data availability (see table A.2).

**Table A.1** Aridity Index ranges used to define dryland zones

| Aridity Class | Definition | Aridity Index range |
|---|---|---|
| 1 | Hyper-arid | 0.00–0.03 |
| 2 | Hyper-arid | 0.03–0.05 |
| 3 | Arid | 0.05–0.20 |
| 4 | Dry semi-arid | 0.20–0.35 |
| 5 | Wet semi-arid | 0.20–0.50 |
| 6 | Dry subhumid | 0.50–0.65 |

**Table A.2** Coverage of the different modeling approaches

| Region | Country | Included in | | |
|---|---|---|---|---|
| | | Irrigation model | Crop model | Livestock model |
| East Africa | Djibouti | ✓ | | |
| | Eritrea | ✓ | | |
| | Ethiopia | ✓ | ✓ | ✓ |
| | Kenya | ✓ | ✓ | ✓ |
| | Somalia | ✓ | | |
| | South Sudan | | | |
| | Sudan | ✓ | | |
| | Uganda | ✓ | ✓ | ✓ |
| | Tanzania | ✓ | ✓ | |
| West Africa | Benin | ✓ | ✓ | |
| | Burkina Faso | ✓ | ✓ | ✓ |
| | Chad | ✓ | ✓ | ✓ |
| | Côte d'Ivoire | ✓ | ✓ | |
| | Gambia, The | ✓ | ✓ | |
| | Ghana | ✓ | ✓ | |
| | Guinea | ✓ | | |
| | Guinea-Bissau | ✓ | | |
| | Liberia | ✓ | | |
| | Mali | ✓ | ✓ | ✓ |
| | Mauritania | ✓ | ✓ | ✓ |
| | Niger | ✓ | ✓ | ✓ |
| | Nigeria | ✓ | ✓ | ✓ |
| | Senegal | ✓ | ✓ | ✓ |
| | Sierra Leone | ✓ | | |
| | Togo | ✓ | ✓ | |
| Central Africa | Burundi | ✓ | | |
| | Cameroon | ✓ | | |
| | Central African Republic | ✓ | | |
| | Congo, Rep. | ✓ | | |
| | Congo, Dem. Rep. | ✓ | | |
| | Equatorial Guinea | ✓ | | |
| | Gabon | ✓ | | |
| | Rwanda | ✓ | | |
| Southern Africa | Angola | ✓ | | |
| | Botswana | ✓ | | |
| | Lesotho | ✓ | | |
| | Madagascar | ✓ | | |
| | Malawi | ✓ | | |
| | Mozambique | ✓ | | |
| | Namibia | ✓ | | |
| | South Africa | ✓ | | |
| | Swaziland | ✓ | | |
| | Zambia | ✓ | | |
| | Zimbabwe | ✓ | | |

The data required for the overall population projections were available for almost all countries in Sub-Saharan Africa.

The data required for the vulnerability analysis were not available for all countries. For East and West Africa, the two subregions on which the analysis concentrates, the coverage was quite limited for East Africa and much more complete for West Africa.

The data required for the resilience analysis similarly were not available for all countries, although the extent of coverage varied depending on the intervention:

- Irrigation development: Data were available for all countries.
- Rainfed cropping systems: Data were available for most of the countries classified as dryland countries.
- Livestock systems: Data were available only for a subset of dryland countries.

The coverage of the overall resilience modeling analysis is thus defined by the coverage of the livestock systems model, which is the narrowest among the various components. The countries included in the overall resilience analysis account for 85 percent of the projected 2030 population in West Africa and nearly 70 percent of the projected population in East Africa (figure A.1).

## Estimation of 2010 baseline population

As discussed at length in the main text of the book, for purposes of the Africa Drylands Study, resilience is determined by three key factors: (1) exposure to droughts and other shocks, (2) sensitivity to droughts and other shocks, and

**Figure A.1** Coverage of the umbrella model: Drylands population equivalent of countries included in the analysis

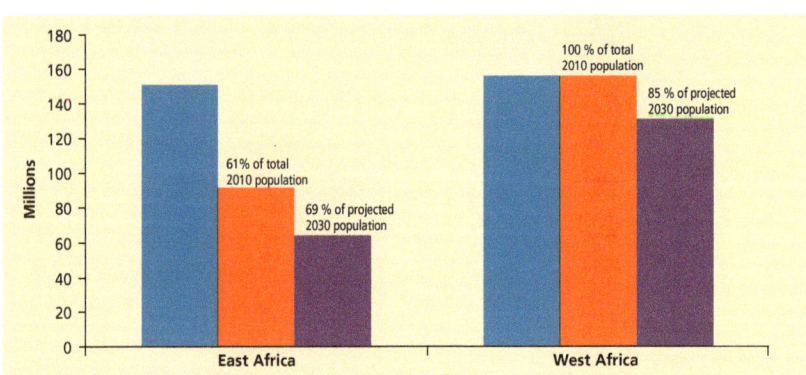

(3) ability to cope with the effects of droughts and other shocks. The estimation of the 2010 baseline population was designed to generate estimates of the numbers of people falling into each of the three categories.

## People exposed to droughts and other shocks

These are defined as people living in dryland areas, that is, areas with aridity classes ranging from hyper-arid to dry subhumid. UN population data were spatialized using gridding methods routinely used in the literature (in particular the Global-Urban Mapping Project [GRUMP] dataset developed at the Center for International Earth Science Information Network—CIESIN—at Columbia University).

## People sensitive to droughts and other shocks

This group is defined as the share of people dependent on agriculture, estimated using recent IMF (International Monetary Fund) estimates (Fox et al. 2013) of the employment shares of agriculture, and assuming that people below working age depend on agriculture in the same proportion as people above working age. All those working in agriculture are assumed to be equally sensitive to drought shocks. This is admittedly a simplification, since the income share derived from agriculture varies across households. However, data needed to assess consistently across countries the income shares derived from agriculture are not readily available. Survey-based evidence suggests, however, that in dryland areas the share of income coming from farming and livestock is at least 60 percent of the total, so this assumption should not excessively bias the analysis.

## People unable to cope with the effects of droughts and other shocks

This group is defined as the proportion of exposed and sensitive people living below the international poverty line of US$1.25 per day. Since separate estimates are rarely available for the rural population only, the national poverty rate was used. The resulting estimates of the number of vulnerable people are probably conservative, because: (1) poverty is usually higher in rural areas than in urban areas, and (2) poverty is usually higher in dryland areas than in non-dryland areas.

Recognizing that in drought years, people dependent on agriculture experience income losses, in some of the analyses in this book the number of people unable to cope is estimated using different poverty lines. Based on survey evidence from the United Nations World Food Programme (WFP), it is assumed that households with incomes exceeding the international poverty line of US$1.25 per person per day by 15 percent, 30 percent, and 45 percent become unable to cope in the event of mild, moderate, and severe droughts, respectively. In each case, the corresponding poverty headcount is estimated based on income distribution data obtained from the PovCalnet database.

Using these definitions, the dimensions of vulnerability and resilience in the drylands of Sub-Saharan Africa were estimated in the baseline year of 2010.

## Resilience analysis for livestock systems

Five simulation models were used to estimate the likely impacts of resilience-enhancing interventions on feed balances, livestock production, and household income resilience, under different climate scenarios (baseline, mild drought, severe drought).

1. The *BIOGENERATOR* model developed by *Action Contre la Faim* (ACF) uses NDVI (Normalized Difference Vegetation Index) and DMP (Dry Matter Productivity) data collected since 1998 from Spot 4 and 5 (Ham and Filliol 2012). The model was used to estimate spatially referenced usable biomass (that is, biomass that is edible by livestock) in the drylands.

2. The *Global Livestock Environmental Assessment Model—GLEAM* developed by Gerber et al. (2013) calculates at pixel and aggregate level: (1) crop byproducts and usable crop residues; (2) livestock rations for the different types of animals and production systems, assuming animal requirements are first met by high-value feed components (crop byproducts if given, and crop residues), and then by natural vegetation; (3) feed balances at pixel and aggregate level, assuming no mobility at pixel level and full mobility at grazing shed level; and (4) greenhouse gases (GHG) emission intensity.

3. On the basis of the feed rations provided by GLEAM, the *IMPACT model* developed by the International Food Policy Research Institute (IFPRI) was used to calculate the production in drylands of meat and milk and to estimate how production will affect overall supply of and demand for these products in the region.

4. The *CIRAD/MMAGE model* consists of a set of functions for simulating dynamics and production of animal or human populations that are categorized by sex and age class. The CIRAD/MMAGE model was used to calculate the sex/age distribution of the four main ruminant species (cattle, camels, sheep, and goats), the feed requirements in dry matter, and milk and meat production.

5. The *ECO-RUM model* developed by CIRAD under the umbrella of the African Livestock Platform (ALive) is an Excel-supported herd dynamics model based on the earlier ILRI/CIRAD DYNMOD. The ECO-RUM model was used to estimate the socioeconomic effects of changes in the technical parameters of the flock or herd (e.g., return on investments, income, and contribution to food security). The modeling exercise benefitted from livestock distribution data contained in the Gridded Livestock of the World

(GLW) database (Wint and Robinson 2007) and its most recent update GLW 2.0 (Robinson et al. 2014). The analysis was informed as well by information and analysis produced by the FAO *livestock supply/demand model* (Robinson and Pozzi 2011). For details, see De Haan (2016).

The results of the models were used as inputs into the final step of the analysis, namely the assessment of the number of households falling into each of three categories: (1) resilient, (2) vulnerable to shocks, and (3) likely to move out of livestock-based livelihoods. These three groups were estimated based on their ownership of livestock, measured in terms of Tropical Livestock Units (TLU). The values of the thresholds used to classify households into one of the three categories were estimated using ECO-RUM, and the corresponding population shares were calculated using a log-normal estimate of the TLU distribution, which approximates quite well (figure A.2) actual TLU distributions emerging from survey data (Survey-based Harmonized Indicators Program [SHIP] database).

The share of households $p_t$ estimated to own less than a certain TLU threshold $t$ is estimated as follows:

$$p_t = \int_0^t f(\tau,\mu,\sigma)d\tau$$

**Figure A.2** Burkina Faso: Cumulative distribution of cattle ownership

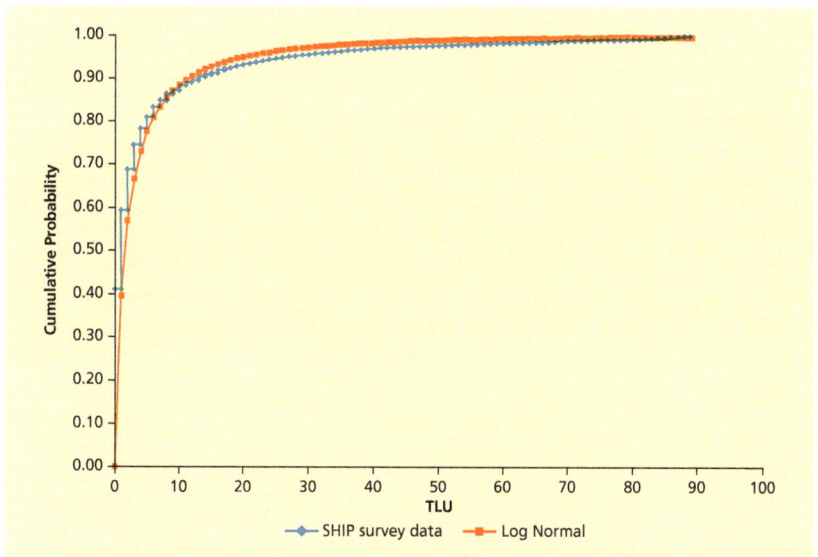

where $f(\tau, \mu, \sigma)$ is the lognormal probability distribution function; and the two parameters $\sigma$ and $\mu$ are estimated as follows:

$$\sigma = \sqrt{2}\ \Phi^{-1}\left(\tfrac{G+1}{2}\right)$$

where $\Phi^{-1}\left(\tfrac{G+1}{2}\right)$ is the inverse of the standard cumulative normal distribution; $G$ is the Gini coefficient, calculated from SHIP survey data (table A.3); and:

$$\mu = \ln(\bar{t}) - \frac{\sigma^2}{2}$$

where $\bar{t}$ is the average number of TLU/household, calculated by dividing the estimate of the total TLU for the relevant country/production system by the corresponding estimated number of households.

Details on the TLU estimates by country and livestock production systems are contained in the background paper on livestock prepared for this study (De Haan 2016).

The critical TLU thresholds are as follows:

- Below 5 TLU per household: households are assumed to feel pressure to drop out of pastoralism.
- 5–19 TLU per household: households are assumed to continue as pastoralists, but are expected to be vulnerable to drought and other shocks.
- Above 19 TLU per household: households are assumed to be resilient to drought and other shocks.

**Table A.3** Gini coefficient of livestock ownership

| Country | Survey year | Income Gini | Livestock Gini | Notes |
|---|---|---|---|---|
| Burkina Faso | 2003 | 39.60 | 52.07 | Survey did not include medium-size livestock |
| Chad | 2011 | 39.78 | 73.99 | Source: Troisieme Enquete sur la Consommation et le Secteur Informel |
| Ethiopia | 2011 | 33.60 | 55.42 | |
| Kenya | 2005 | 47.68 | 78.13 | Excluded TLU > 2,000 (considered outliers) |
| Mali | 2010 | 33.02 | 57.81 | Estimated based on Income Gini |
| Mauritania | 2008 | 40.46 | 66.49 | Estimated based on Income Gini |
| Niger | 2007 | 43.89 | 67.26 | |
| Nigeria | 2004 | 42.93 | 76.63 | Excluded TLU > 1,500 (considered outliers) |
| Senegal | 2005 | 39.19 | 76.05 | |
| Tanzania | 2007 | 37.58 | 67.32 | Survey did not include medium size livestock; excluded TLU >5,000 (outliers) |
| Uganda | 2005 | 42.62 | 54.70 | Calculation only includes medium-size livestock (figures on large-size livestock appear dubious) |

In addition to the Gini coefficient (which is assumed constant throughout the simulation, with the exception of parametric reductions used to simulate the effect of redistribution policies), the other key parameter that determines the number of households below or above the thresholds is the average number of TLU/household.

The average number of TLU/household is estimated by dividing the total number of TLU in the drylands by the total number of households. The numerator in this expression is the maximum number of TLU that the existing biomass can support (on average), estimated through feed balance and herd modeling, based on different levels of access to feed as determined by herd mobility, access to water, insecurity, and urban and crop expansion (further details are provided in De Haan 2016). The denominator in the expression is the number of households estimated to be living in the drylands, based on population growth and projected economic transformation (as explained elsewhere in the book).

The effect of the livestock interventions on vulnerability (and thus indirectly on the number of drought-affected people) is captured by running the model with different values of the TLU resilience threshold (table A.4), estimated through ECO-RUM herd modeling. Lower TLU thresholds imply that for a given distribution of livestock assets, more households will be above the threshold, and fewer households will be below the threshold, compared to the business as usual/no intervention scenario.

Interventions that result in the improvement of animal health reduce the mortality rate and increase the number of animals that can be sold, thereby reducing the number of TLU needed to reach a certain level of income (in particular, the international poverty line of US$1.25/day). Similarly, interventions that promote the sale of animals at a younger age for fattening in high rainfall areas increase the price received per animal and reduce overall mortality, similarly reducing the number of TLU needed to reach a certain income level.

## Resilience analysis for rainfed cropping systems

Similarly to the case of livestock, potential impacts on resilience of interventions targeting rainfed cropping systems are modeled. The analysis is carried out in two stages. In the first stage the objective is to estimate the potential impact of

**Table A.4** Tropical Livestock Units (TLU) required to attain resilience

| Livestock system | Business as usual | | | Health and early offtake | | |
|---|---|---|---|---|---|---|
| | Baseline weather | Mild drought | Severe drought | Baseline weather | Mild drought | Severe drought |
| Pastoral | 21.1 | 23.3 | 24.8 | 15.7 | 17.4 | 18.7 |
| Agro-pastoral | 12.9 | 14.2 | 15.3 | 7.4 | 8.3 | 8.5 |

the adoption of best-bet crop farming technologies on the yields of crops grown by agro-pastoralist and crop farming households. In the second stage the objective is to estimate how these yield changes are likely to translate into income changes and how these income changes impact agro-pastoralist and crop farming households.

## Modeling impacts of best-bet technologies on crop yields

The potential impact of the adoption of best-bet crop farming technologies on the yields of crops grown by agro-pastoralist and crop farming households is estimated using IFPRI's grid-based crop modeling platform. Because it would have been impractical to model the full range of crops grown in the drylands, the analysis is carried out using the dominant cereal crop grown in any given location, identified with the help of IFPRI's Spatial Production Allocation Model 2005 (You et al. 2014) in 2,294 grid cells distributed across 16 countries. The dominant rainfed crops are millet and sorghum in arid and dry semi-arid zones, and maize in wet, semi-arid, and dry subhumid zones.

The crop yield simulations were carried out using three crop models that are part of the (Decision Support System for Agrotechnology Transfer) DSSAT Cropping System Model v4.5 (CERES-Maize, CERES-Sorghum, and CERES-Millet). Yields were simulated at the level of each grid cell over a 25-year period. Using the assumption that weather in the drylands during the next 25 years will not be significantly different from weather experienced during the past 25 years, daily weather data 1984–2008 were used as input (Ruane, Goldberg, and Chryssanthacopoulos 2015). Soil properties in each grid cell were represented using IFPRI's HC27 Generic Soil Profiles Database (Koo and Dimes 2013). Planting date windows for the three representative crops were synchronized with the cropping calendar of the ARV model (described below). A representative variety of each crop was selected and used across the region. Additional details on the modeling platform setup are available in Rosegrant et al. (2014).

## Best-bet crop farming technologies

The DSSAT framework was used to assess the potential impact on yields likely to result from the adoption of five best-bet crop farming technologies: (1) drought-tolerant varieties, (2) heat-tolerant varieties, (3) additional fertilizer, (4) agroforestry practices, and (5) water harvesting techniques. The potential impact on yields was modeled separately for each technology, as well as for several combinations of technologies expected to have synergies (e.g., varieties with drought tolerance and heat tolerance, drought- or heat-tolerant varieties grown with additional fertilizer, and drought- or heat-tolerant varieties grown in combination with agroforestry).

## 1. Drought-tolerant varieties

To simulate the likely impacts of adoption of drought-tolerant varieties, which are known to have superior rooting ability in the presence of low levels of soil moisture, the model was adjusted by increasing the soil root growth factor parameter in each soil layer. Enhanced water extraction capability was also simulated by lowering the lower limit parameter in the soil profile. In the case of maize, the sensitivity was reduced by the anthesis-silking interval (ASI) to soil moisture content.

## 2. Heat-tolerant varieties

The species characteristics definition for each of the three indicator crops includes parameters regarding the response of plant growth and grain filling rates to temperature. In the case of maize, for example, the CERES-Maize model defines the optimum and maximum temperatures for grain filling as 27°C and 35°C, respectively. To mimic the ability of heat-tolerant varieties to continue growing and filling grain at higher temperatures, the values of these two parameters were increased by 2°C for the heat-tolerance simulations.

## 3. Additional fertilizer

The baseline, no-intervention scenario includes an inorganic nitrogen fertilizer application rate that is specific to each region, input system, and crop, which was obtained by calibration of simulated raw yields to FAOSTAT-reported country-level yields. For the best-bet fertilizer intervention, the baseline fertilizer application rate was increased by 50 percent.

## 4. Agroforestry

To simulate the improvements in soil fertility expected to result from decomposing leaves from *Faidherbia* trees planted in the same field as the indicator crops, for each cropping cycle an additional input of organic soil amendments was implemented 10 days before planting. The trees were assumed to be 20 years old in year 1, so that the amount of organic matter contributed throughout the simulation period remains constant. Each tree is assumed to produce 100 kg of leaves, of which 4.3 percent is nitrogen. These values are taken from scientific studies in West Africa. Two tree density values were simulated (5 trees per hectare and 10 trees per hectare), to test the sensitivity of crop yields to tree density. Canopy coverage, which determines the area within each field that actually benefits from the decomposition of tree-contributed organic matter, is assumed to be 10 percent and 20 percent for tree densities of 5 trees per hectare and 10 trees per hectare, respectively. These densities have been observed in many locations in the semi-arid drylands where farmer-managed natural regeneration (FMNR) is practiced. It is useful to recall, however, that while *Faidherbia* is distributed throughout the drylands of Africa, it will not emerge through regeneration in all locations.

**5. Water harvesting**

To simulate the potential effects of harvesting runoff and storing it *in situ* for use in supplementary irrigation, a two-stage approach was implemented. The model was first run without any water management practices, and the output was analyzed to identify periods during the growing season when yields are constrained by lack of water. These periods represent opportunities for implementing improved water harvesting and supplementary irrigation practices. The simulation results were also used to determine when supplementary irrigation can have the largest impact on yields (e.g., immediately after germination and before flowering), and also to estimate how much of the harvested water would be available from the *in situ* storage. The model was then run again including harvested runoff water in the form of supplementary irrigation.

## Modeling impacts of crop yield gains on vulnerability

In the second stage of the analysis, the objective is to estimate how changes in the mean level and distribution of yields associated with adoption of the best-bet technologies are likely to translate into income changes and how these income changes could impact agro-pastoralist and crop farming households. This analysis was carried out using the Africa RiskView (ARV) model developed by the African Risk Capacity.

The ARV model uses static drought vulnerability profiles of the population in each area unit to measure the impacts of drought under different scenarios. More precisely, the ARV model estimates the proportion of the population that is likely to be affected by drought in the presence of drought of different magnitudes. The frequency, intensity, and duration of drought is measured in terms of deviations of a rainfall-based drought index (WRSI) below a defined benchmark multiplied by a scaling factor that translates negative WRSI deviations into potential household income deviations.

Noteworthy features of the ARV model include the following:

- Three different threshold WRSI deviations allow the definition of three levels of vulnerability: (1) vulnerability to mild drought, (2) vulnerability to medium drought, and (3) vulnerability to severe drought. For each analysis unit, the overall vulnerability profile is calculated based on the percentages of the population vulnerable to each of the three levels of droughts.

- The scaling factor used determines the impact of WRSI deviation on crop yields, which in turn translates into impacts on agricultural income of households.

- The vulnerability profiles are defined based on household survey data, which reveal the extent to which households in a specific area unit are both (1) exposed to drought (defined by their percentage of total income generated by agriculture-related activities) and (2) able (or not) to absorb and recover from income shocks (defined by their ranking on a wealth scale compared to the national poverty rate).

Using the outputs of the DSSAT crop modeling simulations (described in the previous section) as an input instead of WRSI, the ARV model can simulate the impact of drought without and with the best-bet technologies. To avoid potential distortions associated with using yield estimates instead of WRSI values, it is assumed that the differences in crop yields attributable to adoption of the best-bet technologies translate into equivalent differences in agricultural income (in the ARV model, this is tantamount to setting the scaling factor to a value of 1:1). The threshold deviations from WRSI that define mild, medium, and severe drought are therefore adjusted accordingly for the use of DSSAT-based input data.

Specific vulnerability profiles at Admin1 level (the first level of sub-national jurisdiction) are created for 2010 and 2030. The 2030 profiles are based on a number of assumptions about demographic increases, economic growth, and structural transformation (described above) that determine how the number of people below the poverty line and the percentage of people employed in agriculture will change by 2030. Within each Admin 1 level unit (that is, the first sub-national level of administrative jurisdiction), the vulnerability profiles can be broken down further by aridity zone. Vulnerability profiles for 2010 and 2030 under the medium fertility scenario are available for the majority of East and West African countries. As an example, table A.5 shows for Mauritania the vulnerability profiles for 2010 and 2030 for the three drought cases.

**Table A.5** Mauritania: Drought Vulnerability Profile for mild, medium, and severe drought (population, millions)

| Region | Aridity | Mild drought | | Moderate drought | | Severe drought | |
|---|---|---|---|---|---|---|---|
| | | 2010 | 2030 | 2010 | 2030 | 2010 | 2030 |
| Assaba | Arid | 0.101 | 0.141 | 0.122 | 0.170 | 0.140 | 0.196 |
| Brakna | Arid | 0.094 | 0.132 | 0.113 | 0.159 | 0.131 | 0.183 |
| Gorgol | Arid | 0.095 | 0.134 | 0.115 | 0.161 | 0.133 | 0.186 |
| Gorgol | Dry semi-arid | 0.001 | 0.001 | 0.001 | 0.001 | 0.001 | 0.001 |
| Guidimaka | Arid | 0.031 | 0.044 | 0.038 | 0.053 | 0.044 | 0.061 |
| Guidimaka | Dry semi-arid | 0.043 | 0.060 | 0.052 | 0.073 | 0.060 | 0.084 |
| Hodh Ech Chargui | Arid | 0.115 | 0.161 | 0.139 | 0.195 | 0.160 | 0.224 |
| Hodh El Gharbi | Arid | 0.087 | 0.123 | 0.106 | 0.148 | 0.122 | 0.171 |
| Tagant | Arid | 0.021 | 0.029 | 0.025 | 0.035 | 0.029 | 0.041 |
| Trarza | Arid | 0.092 | 0.129 | 0.111 | 0.155 | 0.128 | 0.179 |
| **Total** | | **0.680** | **0.953** | **0.821** | **1.150** | **0.947** | **1.327** |

The definition of mild, medium, and severe drought is kept the same in both the 2010 and 2030 profiles. Furthermore, since the poverty line of US$1.25/day is used in both the 2010 and 2030 vulnerability profile definitions, a comparison of these two baseline profiles (BAU) gives an indication of how economic growth and structural transformation are likely to impact the proportion of the population vulnerable to drought as defined by the ARV model. For example, in Mauritania, even though the share of the poor in total population is projected to decline, the absolute number of people vulnerable to drought will actually increase by some 40 percent.

It is important to note that the definitions of drought associated with the vulnerability profiles—mild, medium, and severe—are not linked to return periods of drought, nor necessarily to the risk of drought occurring in a particular Admin 1 unit. Rather, the terms are related to levels of household income loss resulting from drought events. For this reason, adoption in an Admin 1 level unit of one of the best-bet crop farming technologies does not change the vulnerability profile prevailing in that unit. Rather, the changes in the mean level and distribution of crop yields registered in that unit following the adoption of the technology *affects the impact on incomes* of a mild, medium, or severe drought, and therefore affects the probability of hitting the drought-specific threshold. To capture the impact in 2030 of adopting one or more of the best-bet technologies, it is necessary to maintain the definition of drought in the model (in terms of the benchmark and thresholds) and then to calculate the changes in expected number of people affected by drought, given likely yield projections for the various intervention and non-intervention scenarios.

For example, consider first the non-intervention scenarios and the medium fertility scenario. Assume that the rainfall and the resulting crop yields that can occur in an area in 2010 and 2030 come from the same distribution, that is, that there is no change in climate. The DSSAT model can be used to generate yields for 25 years for each Admin 1 level/aridity zone unit. Assume these 25 values represent a sample from a yield distribution for both 2010 and 2030. These 25 yield values can be imposed on the 2010 and 2030 vulnerability profiles to estimate possible drought-affected populations in those scenarios. Figure A.3 shows the estimated number of drought-affected people in Mauritania using the 25 yield values.[1]

To estimate the impacts of the best-bet crop farming technologies on vulnerable populations, the DSSAT model was used to simulate how the various technologies impact the mean level and distribution of yields. Distributions of the drought-affected population estimated using the yield values from the 25 simulation years for each best-bet technology can be compared to distributions of drought-affected populations estimated under the baseline scenario in which yields do not benefit from the adoption of any of the best-bet technologies. The

**Figure A.3** ARV estimates of drought-affected people in Mauritania expected for each of 25 simulated yield years

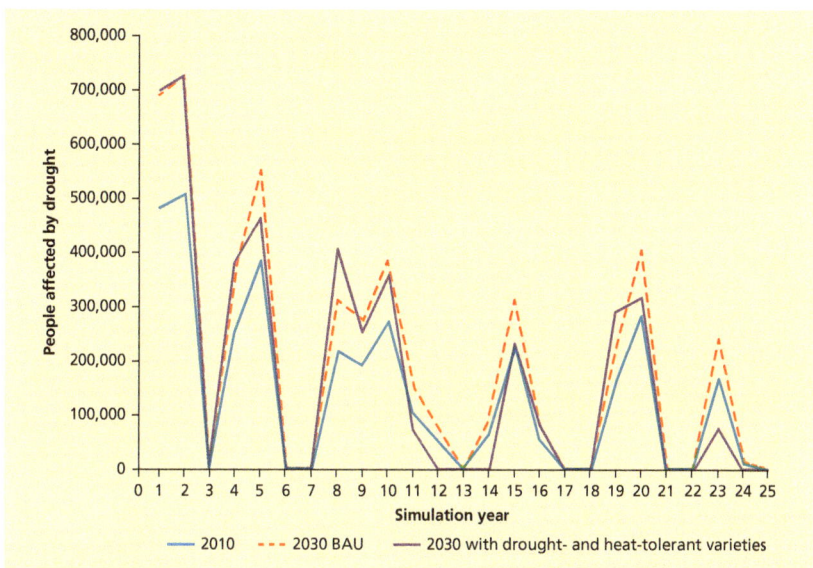

differences show the impact of each technology on the drought-affected population, or in other words, on household resilience.

Figure A.3 shows, again for the case of Mauritania, the effects of adopting one of the best-bet interventions considered in the analysis (specifically, the adoption of a crop variety that is both drought-tolerant and heat-tolerant). Compared to the 2030 no-intervention scenario (BAU), the number of drought-affected people declines in many years; in some years, the result is only to slow down the increase in the number of drought-affected people, while in other years the number of drought-affected people actually falls below the 2010 baseline. Overall, adopting the drought- and heat-tolerant variety leads to an 11 percent decrease in the number of drought-affected people. This example shows the benefit of a single intervention adopted in all polygons where it is effective. In the model, benefits are maximized when the entire set of interventions is considered, and in each polygon the intervention is selected that yields the largest reduction in the number of drought-affected people. The results presented in the main text of the book are based on the latter approach.

## Irrigation resilience analysis

The final intervention modeled is irrigation development. The assessment of the potential impacts of irrigation development on the population living in drylands builds on the same drought characterization method used for the analysis of impacts of interventions in rainfed cropping systems (see the previous appendix section, Resilience analysis for rainfed cropping systems), combined with work done by IFPRI on irrigation investment potential in African drylands (Xie et al. 2015). In the IFPRI work, the potential for expanding large-scale irrigation (LSI) and small-scale irrigation (SSI) in dryland areas of Sub-Saharan Africa by 2030 are modeled separately. (See box A.1 for details on the SSI modeling exercise.)

It is important to note that the area identified as having irrigation investment potential should be interpreted as "physical area equipped with irrigation infrastructure," since the water balance figures used to make the projections are long-term averages. In drought years when water becomes scarce, irrigation can not be delivered everywhere, leaving part of the area equipped with irrigation infrastructure unused. This becomes important in the latter stages of the analysis, when the impacts of irrigation on drought-affected people are estimated in the face of weather variability and climate change.

The impact of irrigation development on reducing vulnerability and increasing resilience in the drylands was assessed using a two-step procedure. The first step involved estimating the area that is actually irrigated, taking into account climatic variability. The second step was to estimate, based on the results of the first step, the population that can be considered no longer affected by drought for each Admin 1 level/aridity zone unit.

The key steps and assumptions used in the analysis are shown below:

SSI can use either surface water or groundwater. Groundwater acts as a buffer against the impact of drought. The abundance of groundwater storage and accessibility to groundwater in African drylands is evaluated through geographic information system (GIS) analysis using groundwater depth and storage data developed by British Geological Survey (table A.6).

**Table A.6** Aquifer classification in British Geological Survey groundwater data

| Class | 1 | 2 | 3 | 4 | 5 | 6 |
|---|---|---|---|---|---|---|
| Depth to groundwater (meters) | 0–7 | 7–25 | 25–50 | 50–100 | 100–250 | >250 |
| Groundwater storage (millimeters) | 0 | <1,000 | 1,000–10,000 | 10,000–25,000 | 25,000–50,000 | >50,000 |

**BOX A.1**

## Estimating the expansion potential for small-scale irrigation (SSI)

The method used to assess SSI development potential begins with an irrigation suitability analysis. Within each pixel, various criteria are used to score the environmental suitability of each pixel, including topography (slope), groundwater accessibility, distance to perennial surface water, proximity to existing irrigation, and market access.

For the ex ante suitability analysis, the criteria parameters are divided into three classes, and linear interpolation is used within the classes to calculate the scores. Such a classification is similar to a stepwise function, which provides flexibility to adjust the threshold values after consulting with experts and stakeholders. The overall rating of the irrigation suitability is the average of all scores for all applicable criteria. Since groundwater and surface water provide the same water resource to irrigation, overall suitability is calculated as the larger of the two scores. In other words:

$$S = \frac{S1 + max\ (S2,S3) + S4 + S5}{4}$$

where: $S$ = irrigation suitability score, $S1$ = score for slope, $S2$ = score for surface water access, $S3$ = score for ground water access, $S4$ = score for ground distance to existing LSI, and $S5$ = score for market access.

The ex ante suitability analysis is done on a 0.5 x 0.5 km grid. The suitability score is then used as a percent of the pixel suitable for irrigation. In other words, the area with SSI development potential in a pixel is calculated as:

$$A_{irr,exante} = A_{pixel}\ x\ \frac{S}{100}$$

where: $A_{irr,exante}$ = area suitable for irrigation development (ha), and $A_{pixel}$ = pixel size (= 25 ha).

Next, the expansion of SSI is simulated. The starting point for the analysis is the current cropping pattern. Data on area harvested, production, and yield under irrigated and rainfed systems on a grid of approximately 10 x 10 km were obtained from the IFPRI Spatial Production Allocation Model (SPAM) database (for details, see You, Wood, and Wood-Sichra 2009). Prior to the simulation, the results of the ex ante suitability analysis were incorporated into the SPAM grid. The suitability score for each SPAM pixel was calculated as the average of the pixels in the coarser grid used for the suitability analysis, and the total area within each SPAM pixel deemed suitable for irrigation was calculated as the sum of the areas within the pixels used for the suitability analysis.

To account for the expansion in cultivated area and changes in cropping patterns that may be caused by irrigation development, the following key assumptions were made:

*(continued next page)*

**Box A.1** *(continued)*

- Irrigation can occur during both the wet and dry seasons (both seasons are recognized in the analysis). Based on empirical evidence from past studies (Xie et al. 2015), the following 10 crops can be irrigated during the rainy season: (1) wheat, (2) rice, (3) maize, (4) sorghum, (5) millet, (6) potatoes, (7) sweet potatoes, (8) groundnuts, (9) sugarcane, and (10) vegetables. Wheat, maize, rice, and vegetables are assumed to be the dry-season irrigated crops.

- During the irrigation expansion, (1) the currently existing rainfed cultivated area in a country will first be converted to irrigated area before new area is brought into cultivation/irrigation; (2) irrigation will expand according to the overall rating of the irrigation suitability, that is, irrigation development first takes place in the pixels with the highest suitability scores and is followed by development in pixels with the second highest ranking; and (3) irrigation expansion is constrained by water availability and national-level food demand for irrigated crops.

The detailed simulation algorithm is described in Xie et al. (2015). It is assumed that the area cultivated for a given crop c on irrigated land, either converted from existing rainfed land or expanded from non-farming area, is proportional to the profitability of cultivating that crop.

$$A_c^i = A_{total} \times \frac{profit_c}{\Sigma_c\, profit_c}$$

where: $A_{total}$ = total irrigated area (ha), and *profit_c* = annual profit farmers receive from cultivating crop c ($/ha).

Profit is calculated as follows:

$$Profit_c = Y_c^i \bullet P_c \bullet ProfitRatio_c$$

where: $Y_c^i$ = yield of crop under irrigation (ton/ha), derived from FAO's Global Agro-Ecological Zones (GAEZ) database (http://www.fao.org/nr/gaez/en/) under an assumption that irrigated yields would be 50 percent of the GAEZ potential yields for the 2050 analysis; for 2030 it is assumed that 80 percent of the 2050 yields can be achieved; $P_c$ = producer price of crop c ($/ton), derived from the FAO PriceSTAT database, *ProfitRatio_c* = profit margin (0 ~1) of crop c (PriceSTAT appendix table 1.3).

To calculate the internal rate of return (IRR), first annual net revenues from the irrigation expansion are calculated without taking into consideration irrigation costs.

The net revenue in the rainy season on converted rainfed land in a SPAM pixel ($/yr) is calculated as:

$$NetRevenue_{wet} = \Sigma_c\, Y_c^i \bullet P_c \bullet ProfitRatio_c \bullet A_c^i - \Sigma_c\, Y_c^r \bullet P_c \bullet ProfitRatio_c \bullet A_c^r$$

where: $Y_c^r$ = rainfed yield of crop c (ton/ha) and $A_c^r$ = rainfed area of crop c in the pixel (ha).

The net revenue in the rainy season on newly cultivated, irrigated land in a SPAM pixel ($/yr) is calculated as:

*(continued next page)*

**Box A.1** *(continued)*

$$NetRevenue_{wet} = \Sigma_c \, Y_c^i \cdot P_c \cdot ProfitRatio_c \cdot A_c^i$$

The net revenue on converted rainfed land or newly cultivated irrigated land in dry season is calculated as:

$$NetRevenue_{dry} = \Sigma_c \, Y_c^i \cdot P_c \cdot ProfitRatio_c \cdot A_c^i$$

The net revenue per unit area (without consideration of irrigation costs) is calculated as:

$$NetRevenue\_per\_ha = \frac{NetRevenue_{wet} + NetRevenue_{dry}}{\Sigma_c \, A_c^i}$$

With the calculated net revenue per unit area, the cash flow in year *t* required for the IRR calculation is calculated as:

$$NetRevenue\_per\_ha * B_t - A_t * IRR\_Cost_c * C_t - IRR\_COST_O$$

where: $IRR\_Cost_c$ ($/ha) = the annualized capital investment cost for SSI expansion, $IRR\_Cost_c$ = annual SSI operating costs ($/ha), and $B_t$ and $C_t$ = factors used to amortize capital investment and revenue in the calculation of cash flow associated with irrigation infrastructure development. The calculation assumes a five-year investment cycle and a 50-year investment horizon.

It is assumed that SSI in areas with groundwater depth below 25 meters (m) and storage greater than 10,000 millimeters (mm) is primarily groundwater-based and not influenced by drought.

The variation of actual area under surface water-based SSI and LSI is modeled as a function of the drought index *I*.

$$A_i = A_O \cdot e^{-\alpha I}$$

where $A_i$ is actual area of irrigation in year *i*; $A_O$ is physical area equipped with irrigation; *I* is the drought index. Its value may vary between 0 and 1. $A_i = A_O$ if $I = 0$; and in a drought year, $I > 0$ and $A_i < AO$, $\alpha$ is a parameter controlling the contraction rate of irrigation area under drought. The higher the value of $\alpha$, the larger the reduction in irrigation area in drought years.

The drought index is calculated as follows:

$$I = \frac{Y_{benchmark} - Y_i}{Y_{benchmark}} \quad if = Y_{benchmark} > Y_i \, ; \; otherwise \; I=0$$

where:

$Y_{benchmark}$ is the benchmark yield defined in the ARV model, and $Y_i$ is the crop yield in a given year *t*. Given that large reservoirs likely have multi-year storage capacity, LSI tends to be more resilient to drought than surface-water-based SSI.

Therefore, a smaller value of $\alpha$ is specified for LSI in the simulation. $\alpha$ is set to 0.5 for LSI and 1.0 for SSI.

The simulation of "actual" LSI and SSI irrigated areas is conducted at 5-arc minute resolution (approximately 10 km by 10 km). The calculated pixel-wise values of "actual" areas of irrigation are aggregated to the Admin 1 level/aridity zone unit. The number of poor people in each unit is calculated under the assumption that "0.5 hectares of irrigated land supports one household (HH) comprising 5 people" and accordingly vulnerability shares are developed from the ARV model as:

$$Pop_i = A_i \ x \ 10 \ x \ \eta$$

where $Pop_i$ is population in a unit and in year $i$ is rendered resilient to drought through irrigation, $A_i$ is actual area of irrigation in the unit, and year $i$, $\eta$ is the vulnerability share of population obtained from the ARV model. A key assumption underlying the analysis is that where there is potential for irrigation development, vulnerable people will be able to take advantage of the opportunity and equip their farm with SSI equipment, regardless of their income level. In other words, the ability to take advantage of opportunities to invest in irrigation is assumed to be the same for every household located in areas with irrigation development potential, irrespective of their income level.

## Consolidating the results of the resilience analysis

Estimated reductions in the numbers of drought-affected people likely to result from interventions in livestock systems and rainfed cropping systems, as well as from investments in irrigation, are consolidated in a set of figures presented in the book.

Key elements of the consolidation process include the following:

- The livestock model was used to generate estimates of the number of vulnerable people (without and with the interventions) in hyper-arid and arid zones only (aridity classes 1 to 3, see figure A.4), using the model's parameters for pastoral livelihoods.
- Results expressed in terms of number of households were converted into numbers of people by assuming an average household size of six people.
- The number of drought-affected people was estimated applying country-specific drought incidence factors (average number of drought-affected people as percentage of vulnerable people) obtained from the crop model. This is justified on account of the likely significant correlation between drought impacts on the staple crops modeled (maize, millet, sorghum) and impacts on the grasses found in rangelands.

**Figure A.4** Schematic of livelihood modeling

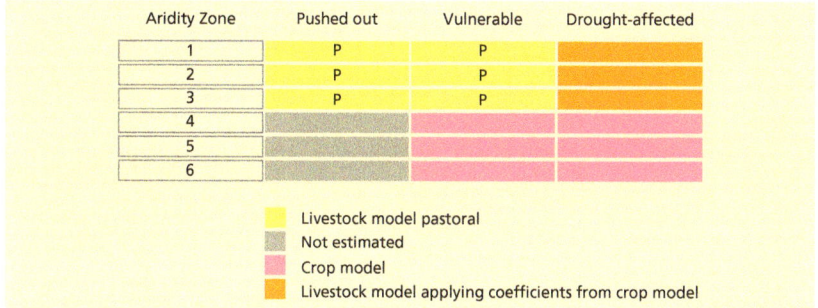

- The livestock model estimates of the number of households below the critical threshold of 5 TLU/household (figure A.4) were used to calculate the number of people who are likely to transition from pastoralism to farming; these households were then added to the number of vulnerable people engaged in crop farming. Country-level estimates of the number of people who are likely to transition from pastoralism to farming were distributed across the country's polygons (intersection of administrative units and aridity zones) using each polygon's share in the country's total number of vulnerable people.
- The number of drought-affected people engaged in crop farming in aridity classes 4 to 6 (including both the original crop farmers as well as the people who are likely to transition from pastoralism to farming) was estimated using the crop model.

The approach used in this book does not consider the significant scope for implementing livestock-related interventions in agro-pastoral systems found in semi-arid and dry subhumid zones. For this reason, while the modeling results indicate the order of magnitude of the likely resilience benefits of the different interventions, they represent conservative lower bound estimates of the full potential.

## Cost estimates

### Livestock
Cost estimates for the analysis of livestock systems are based on cost projections from five recently launched internationally funded projects dealing with pastoral areas.[2] These data were complemented with data obtained through a review of the literature. Table A.7 provides a summary of the cost per pastoral/agro-pastoral person associated with these projects.

**Table A.7** Average cost/person/year (weighted according to number of beneficiaries) of the main interventions in five dryland livestock development projects

| Intervention | Average cost/person/year (US$) | Number of projects | Range (US$) |
|---|---|---|---|
| Health improvement | 3.95 | 3 | 3.37–20.12 |
| Market improvement (early offtake of bulls) | 6.00 | 3 | 3.67–8.33 |
| Early warning systems | 3.72 | 2 | 1.79–2.09 |
| Social services, etc. | 5.30 | 2 | 2.39–5.82 |

The range of values is significant, particularly for health improvement. However, the average is in line with the estimates of the World Organisation for Animal Health (OIE)-sponsored study (CIVIC Consulting 2009) for Uganda.

For development decision making, it is important to know the distribution between technology adoption-related and non-adoption-related costs, as well as between investment and recurrent costs. The assumptions used are based on the projects analyzed and the authors' experience; they are provided in table A.8.

**Table A.8** Assumptions about the allocation of adoption- and non-adoption-related costs and of investments and recurrent costs for animal health and early offtake interventions

| Item | Allocation |
|---|---|
| Animal health non-adoption-related | Of total health improvement budget, 20% in investments and 25% in recurrent costs |
| Animal health adoption-related | Of total health improvement budget, 25% in investment and 30% in recurrent costs |
| Animal health improvement adoption-related by livestock system | 10% higher/person (higher delivery costs) in pastoral systems |
| Early offtake (market integration) | Of total budget, 70% in investment and 30% in recurrent costs (high capital investment needed in infrastructure such as transport, processing facilities) |
| Early offtake non-adoption-related costs | Nil, because of its currently nascent character |
| Adoption rate | 70% for pastoral and 80% for agro-pastoral households for health improvement and 60% and 70%, respectively, for early offtake |
| Public and private sector contribution | Public sector: 80% for cross-cutting costs, 60% for adoption costs in animal health improvement, and 20% for early offtake; the remainder belongs in the private sector |

In aggregate, these figures seem high, at a total of about US$10 billion over the 20-year period (table A.9) or about US$500 million/year (about US$200 million/year for the public sector).

They look more reasonable when calculated per beneficiary (number of people made resilient), as shown in figure A.5.

**Table A.9** Summary of costs (2011–14 prices, US$ billion) of health and early offtake interventions and their distribution between the public and private sectors (2011–30)

|  | Cross-cutting costs | Adoption costs animal health | Early offtake costs | Total |
|---|---|---|---|---|
| Public sector | 1.14 | 1.69 | 1.18 | 4.01 |
| Private sector | 0.29 | 1.13 | 4.71 | 6.12 |
| Total | 1.43 | 2.82 | 5.88 | 10.14 |

Figure A.5 shows that with the exception of Niger, the costs per person made resilient are significantly below the US$100–135 normally calculated for food aid. As expected, the annual cost per person made resilient is higher in pastoral areas. In general, the costs in East Africa seem to be lower than in the Sahel. At an average cost of US$27 per person per year, they are half the US$65 per person per year estimated by Venton et al. (2012).[3]

## Rainfed crops

The cost of adopting the rainfed cropping technologies includes public costs borne by the public sector during an initial period when a technology is first being introduced (e.g., costs associated with extension campaigns, demonstrations, free samples; see table A.10), as well as private costs borne by the adopting farmers themselves (e.g., the cost of purchasing seed or fertilizer, or the cost of performing additional operations such as planting fertilizer trees or building water harvesting structures).

**Figure A.5** Estimated unit cost (US$/person made resilient/year, expressed on a log scale) under baseline climate and health and early offtake scenarios

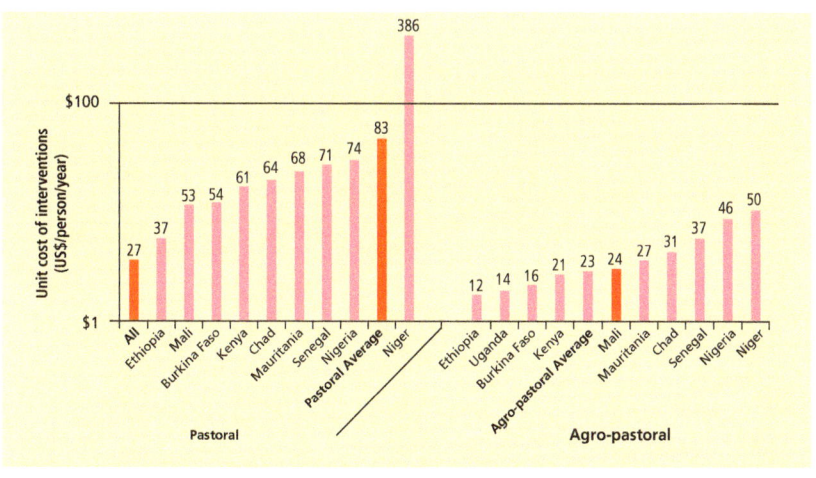

**Table A.10** Public costs of technology transfer (US$/hectare)

| Description | Millet | Sorghum | Maize |
|---|---|---|---|
| 1: Drought tolerance | 1.25 | 1.35 | 1.50 |
| 2: Heat tolerance | 1.25 | 1.35 | 1.50 |
| 3: More fertilizer | 10.00 | 10.00 | 10.00 |
| 4_5: Agroforestry 5 trees/ha | 45.00 | 45.00 | 45.00 |
| 4_10: Agroforestry 10 trees/ha | 45.00 | 45.00 | 45.00 |
| 5: Water harvesting | 20.00 | 20.00 | 20.00 |

Private costs (that is, costs borne by farmers themselves) were included in the analysis by adjusting downward the yield gain associated with adoption of the technology by a discount factor estimated to represent the cost of adopting the technology. To reflect the fact that farm households will use part of their income to purchase the inputs required for adopting the technology (e.g., labor, seed, fertilizer), costs were expressed in terms of the crop equivalent of purchasing the required inputs, with production valued at country- and crop-specific farm-gate prices calculated as averages of the corresponding FAOSTAT values over the period 2000–12.

The cost (estimated on the basis of the literature and expert judgment) varied by technology (table A.11). In some cases it was modest (e.g., adoption of drought-tolerant and heat-tolerant varieties, adoption of FMNR), whereas in other cases it was more substantial (e.g., additional fertilizer, water harvesting). In recognition that technology adoption costs may be borne by the farmer or by the state (in the form of subsidies), sensitivity analysis was carried out to explore the impacts on adoption incentives of differing levels of private costs.

To reflect the fact that the best-bet crop farming technologies will not all be profitable in every location, a switch was built into the model to determine which technology is adopted in any given polygon. The switch works as follows: if adoption of a given best-bet technology has the effect of reducing the number of drought-affected people, that technology is deemed effective and retained,

**Table A.11** Private costs of technology adoption (US$/hectare)

| Description | Millet | Sorghum | Maize |
|---|---|---|---|
| 1: Drought tolerance | 3 | 3 | 15 |
| 2: Heat tolerance | 3 | 3 | 15 |
| 3: More fertilizer | 30 | 30 | 30 |
| 4_5: Agroforestry 5 trees/ha | 7 | 7 | 7 |
| 4_10: Agroforestry 10 trees/ha | 9 | 9 | 9 |
| 5: Water harvesting | 45 | 45 | 45 |

but if adoption of that technology has the effect of increasing the number of drought-affected people, the technology is deemed ineffective and discarded.

In addition, because synergies resulting from the simultaneous adoption of multiple best-bet technologies are not captured well by the DSSAT model, the analysis used the simplifying assumption that only the most effective technology is adopted in a given location. Because simultaneous adoption of multiple technologies would certainly result in additional benefits (in terms of yield increases and income gains), the resilience-enhancing impacts of adoption of improved rainfed cropping technologies should be considered conservative.

## Irrigation

Given the considerable uncertainty and wide range of irrigation technology and expansion costs, three sets of cost assumptions were considered in the analysis of irrigation development, ranging from US$8,000–US$30,000 per hectare for LSI, and from US$3,000–US$6,000 per hectare for SSI (table A.12). The medium-cost assumptions were used for the baseline scenario.

**Table A.12** Irrigation development unit cost assumptions (US$/hectare)

|  | Low | | Medium | | High | |
|---|---|---|---|---|---|---|
|  | Capital | Operation and maintenance | Capital | Operation and maintenance | Capital | Operation and maintenance |
| LSI | 8,000 | 800 | 12,000 | 1,200 | 30,000 | 3,000 |
| SSI | 3,000 | 100 | 4,500 | 125 | 6,000 | 150 |

*Source:* Xie et al. 2015.

## Notes

1.  The national population affected is the sum of the populations affected in each Admin 1/aridity zone.
2.  The Ethiopia-Drought Resilience & Sustainable Livelihood Program in the Horn of Africa (PHASE I), funded by the African Development Bank (US$48.5 million, 2012); the International Fund for Agricultural Development (IFAD)- and World Bank-funded Regional Pastoral Livelihoods Resilience Project for Kenya and Uganda (US$132 million, 2014); the World Bank-funded Regional Sahel Pastoralism Support Project (US$250 million, under preparation); the World Bank/IFAD-funded Ethiopia Pastoral Community Development Project–Phase II (US$133 million, 2013); and the IFAD-funded Sudan Livestock Marketing and Resilience Program (US$119 million, under preparation).
3.  US$54/person/year for Kenya and US$77/person/year for Ethiopia. No data are available for the Sahel.

## Appendix References

Carfagna, F., R. Cervigni, and P. Fallavier, eds. 2016 (forthcoming). *Mitigating Drought Impacts in Drylands: Quantifying the Potential for Strengthening Crop- and Livestock-Based Livelihoods.* World Bank Studies. Washington, DC: World Bank.

CIVIC Consulting. 2009. "Systems for Animal Diseases and Zoonoses in Developing and Transition Countries." Study sponsored by OIE, World Bank, and European Union. http://www.oie.int/doc/document.php?numrec=3835503.

De Haan, C., ed. 2016. *Improved Crop Productivity for Africa's Drylands.* World Bank Studies. Washington, DC: World Bank.

Fox, L., C. Haines, J. Huerta Muñoz, and A. Tho. 2013. "Africa's Got Work to Do: Employment Prospects in the New Century." IMF working Paper 13/201. International Monetary Fund (IMF), Washington, DC.

Gerber, P.J., H. Steinfeld, B. Henderson, A. Mottet, C. Opio, J. Dijkman, A. Falcucci, and G. Tempio. 2013. *Tackling Climate Change through Livestock: A Global Assessment of Emissions and Mitigation Opportunities.* Rome: Food and Agriculture Organization of the United Nations (FAO).

Ham, F., and E. Filliol. 2012. "Pastoral surveillance system and feed inventory in the Sahel." Chapter 10 in M.B. Coughenour and H.P.S. Makkar (eds), *Conducting National Feed Assessments.* FAO, Rome.

Koo, J., and J. Dimes. 2013. HC27 Generic Soil Profile Database, http://hdl.handle.net/1902.1/20299. International Food Policy Research Institute [Distributor] V2 [Version].

Robinson, J., and F. Pozzi. 2011. "Mapping Supply and Demand for Animal-Source Foods to 2030." Animal Production and Health Working Paper. No. 2. FAO, Rome.

Robinson, T.P., G.R.W. Wint, G. Conchedda, T.P. Van Boeckel, V. Ercoli, E. Palamara, G. Cinardi, L. D'Aietti, S.I. Hay, and M. Gilbert. 2014. "Mapping the Global Distribution of Livestock." *PLoS ONE* 9(5): e96084. doi:10.1371/journal.pone.0096084.

Rosegrant, M.W., J. Koo; N. Cenacchi, C. Ringler, R.D. Robertson, M. Fisher, C.M. Cox, K. Garrett, N.D. Perez, and P. Sabbagh. 2014. "Food Security in a World of Natural Resource Scarcity: The Role of Agricultural Technologies." International Food Policy Research Institute (IFPRI). Washington, DC. http://ebrary.ifpri.org/cdm/ref/collection/p15738coll2/id/128022.

Ruane, A.C., R. Goldberg, and J. Chryssanthacopoulos. 2015. "AgMIP Climate Forcing Datasets for Agricultural Modeling: Merged Products for Gap-Filling and Historical Climate Series Estimation." *Agr. Forest Meteorol.* 200: 233–48, doi:10.1016/j.agrformet.2014.09.016.

Venton, C.C., C. Fitzgibbon, T. Shitarek, L. Coulter, and O. Dooley. 2012. "The Economics of Early Response and Disaster Resilience: Lessons from Kenya and Ethiopia." Economics of Resilience Final Report. https://www.gov.uk/government/uploads/system/uploads/attachment_data/file/67330/Econ-Ear-Rec-Res-Full-Report_20.pdf.

Wint, G.R.W., and T.P. Robinson. 2007. "Gridded Livestock of the World." FAO, Rome.

You, L., S. Wood, and U. Wood-Sichra. 2009. "Generating Plausible Crop Distribution Maps for Sub-Saharan Africa Using a Spatially Disaggregated Data Fusion and Optimization Approach." *Agricultural System* 99(2–3): 126–40.

You, L., U. Wood-Sichra, S. Fritz, Z. Guo, L. See, and J. Koo. 2014. "Spatial Production Allocation Model (SPAM) 2005 Beta Version. 2015." Available from http://mapspam. info.

Xie, H., W. Anderson, N. Perez, C. Ringler, L. You, and N. Cenacchi. 2015. "Agricultural Water Management for the African Drylands South of the Sahara." Background report for the Africa Drylands Study. International Food Policy Research Institute, Washington, DC.

green
press
INITIATIVE